DON'T DIGITISE YOUR RUBBISH

DON'T DIGITISE YOUR RUBBISH

INTEGRATE, SIMPLIFY, AND SYSTEMATISE YOUR OPERATIONS FIRST

ANDY SHERRING

HOUNDSTOOTH
PRESS

DON'T DIGITISE YOUR RUBBISH
Integrate, Simplify, and Systematise Your Operations First

FIRST EDITION

ISBN   978-1-5445-2815-1 *Hardcover*
          978-1-5445-2814-4 *Paperback*
          978-1-5445-2813-7 *Ebook*

CONTENTS

INTRODUCTION 9

1. THE INDUSTRY HAS A SERIOUS PROBLEM
   AND IS IN A VULNERABLE POSITION 19

2. STRUCTURAL WEAKNESSES NEED TO BE
   ADDRESSED 55

3. THE OPERATIONAL BASICS NEED TO BE
   TAKEN TO A WHOLE NEW LEVEL 89

4. INTEGRATION PROVIDES THE MISSING LINK
   FOR TRANSFORMATIONAL CHANGE 161

5. A FIVE-STEP GUIDE FOR INTEGRATING YOUR
   BUSINESS 195

CONCLUSION..229

SUMMARY 235

ACKNOWLEDGMENTS 245

INTRODUCTION

///

The mining industry has a problem. It is facing a period of major technological and social change, which it is not currently ready for. There is huge value at stake for mining companies and for the industry as a whole. The next five to ten years will prove to be pivotal in separating those who proactively prepare well for this change, and those who wait for the change to happen to them before reacting.

For senior decision makers who have the authority to drive transformational change, this book is essential reading. For anyone in, or aspiring to be in, a leadership role within mining or a similar operational industry, you will certainly also benefit and find it to be of great interest.

The mining industry is not unique in its failure to prepare adequately for change. But in over four decades of experience along my journey in operations, I saw issue after issue. I was always struck by the lack of consistency of approach from an operator's perspective. Misalignment and fragmentation of effort, siloed thinking, self-serving behaviour, reinventing the wheel, and a short-term focus were consistently problematic. These, of course, are natural challenges in a large business, and I have grown to see that the same exists in most large operational businesses across the world. Only the degree of dysfunctionality differs.

However, many of these problems just seemed so basic, especially for large and sophisticated corporations. Things like having some level of standardisation across the business, with respect to systems and processes, organisation models, and even operating philosophies. These mostly didn't exist, and when they *were* implemented, they generally fell by the wayside some years later. What this led to was a reactive approach to doing things and a constant reinventing of the wheel every time a new leader arrived. Too often, the new leader would commence in their new role by throwing out everything that the previous manager had done, then starting from scratch, using their own personal past experience. Nothing systemic, nothing sustainable, and nothing that encouraged continuous improvement over time.

It always struck me that there had to be a better way than this. I was internally driven by the challenge to define what that 'better way' actually was. I saw it as a profoundly important issue, yet no one seemed to be interested in cracking this conundrum. Maybe it was put in the 'too hard' basket. In any case, it had somehow just become the accepted norm. The attitude seemed to be, 'Well, that's mining'.

Perhaps the problem, at its core, is that the senior operational leadership of businesses are so caught up in managing short-term challenges today that they simply don't have the headspace to think about long-term, systemic, sustainable change. This is not meant as a criticism, because it is a fact that frustrates most leaders in the industry. But it must be said. It is a clear indication of the vicious circle that these senior leaders currently find themselves stuck in. Over the past two decades, for all sorts of reasons, the focus of business unit CEOs and their teams has shifted from 'long-term shareholder value' to 'short-term share price'. They pursue lots of short-term improvement initiatives and regular organisational structure changes, but somehow nothing that truly turns the dial.

Illustrating this, the mining industry as a whole has seen a substantial decline in productivity over the past twenty years, much more than can be explained away by declining ore grades. And that is in the face of better equipment, better technology, and vastly more data and analytical capability. It is a sad fact that the business improvement goal of most mining companies today is to get back to the productivity levels of fifteen to twenty years ago. Something is wrong with the current approach. It is not working.

The world of operations has changed profoundly over the past twenty years. The 'complexity' of the role of, say, a mine general manager (GM) is completely different today than what it was in the past. And the 'capability' of their management team is typically completely different, due to shorter tenure and much less depth of local knowledge and experience. These are largely just an unavoidable reality of a changing world. However, what is crazy is that businesses have not changed their operating models. The complexity and capability combination has had a double-whammy impact, yet businesses continue to follow the same approach they always have—the same approach to organisational change, operating models, and business improvement that has not delivered systemic, sustainable improvement over the past twenty years.

And now we have the rapid explosion of technological advancement to add to the equation. Everyone knows this is real, and it's coming fast, but anyone with any appreciation for the current realities in operations knows that technology is not a silver-bullet solution. They know they have to embrace the future, but they also recognise that digitising poor operating practices is not the answer. It will just automate those weak practices. Hence, the title of this book: *Don't Digitise Your Rubbish!*

There is a need to fix the basics first. Doing this in the right way will bring greater clarity, focus, alignment, accountability, and discipline to the operations. This book discusses the changes in the industry in detail, and presents a solution: a better way of setting up operations and businesses, which not only addresses the problems of today, but also provides a solid platform for the future.

The key missing link is integration. It is what lies at the heart of most problems facing large operating businesses, and it is also the perfect vehicle for driving the necessary change.

It helps leaders become crystal clear on the business strategy and operating philosophy; then it creates a working environment (both organisational and physical), which aligns everyone and everything within the business around delivery of that strategy. In the process, all other obstacles, clutter, and bureaucracy standing in the way of this are identified and removed. In short, the process makes it easier for everyone to do their jobs effectively, by integrating, aligning, and simplifying everything that is being done.

Implementing a fully integrated operating model across your operations will create a better, more systemically focussed organisation, which is capable of much greater things than it is today. It will position you for better performance in the short term, and it will prepare you well for the digital future. It will liberate substantial latent untapped value, which lies stuck in all the countless interfaces across your organisation. It will pull your teams together around a common drumbeat, which optimises the performance of the operating system as a whole. It will also inspire and enable them to do their jobs more efficiently and effectively.

This book will show you what this involves and how to go about it. It presents a practical solution for setting up operations and businesses for short-, medium-, and long-term success. It is written from a position of operational empathy, not from the perspective of technical theory. It will guide you through the problem, the challenges, and the solution, including: why the industry has a serious problem; why it's in a vulnerable position; what are its structural weaknesses that need to be addressed; what is the detail around the operational basics that need to be taken to a whole new level; why and how integration provides the missing link for transformational change; and it concludes with a five-step guide for integrating your business.

It is structured to make it easy to read, and the key points are made clearly. In each chapter and section of the book, the industry observations are highlighted, the key principles are discussed, and a solution is provided. This is followed by a useful checklist of questions for the reader to consider, with regard to their own business and/or operations. And each section includes a powerful illustration that visually captures the key points from the section.

My background has perfectly prepared me to get to this place where my business and I now specialise solely in integration, and we have developed robust solutions for long-standing industry issues.

I had a twenty-eight-year career with Rio Tinto (a leading global mining group), working in operational roles across the globe, from copper in South Africa, to salt in Australia, to coal in the US, and back to Australia in technology and innovation, and then iron ore. My last role with Rio Tinto was to lead the design and delivery of the groundbreaking and highly successful world-first remote Operations Centre for the iron ore group in Perth. This was a catalyst for me, and I felt we had just cracked the door open into a whole new world of opportunity related to operational excellence, which went far beyond just a 'centre' and was much deeper than just the integrated planning function that accompanied it.

I then went into independent consulting, focussing solely on this subject. I consulted to around twenty-five of the major mining companies around the world, working across a diversity of different commodities, applications, and cultures. Over a period of five years, I began to really understand what this 'better way' actually was. I wanted to know: What are the key principles? Why is this important? Why wasn't it important previously? Why is it so critical an issue right now? What is involved in fixing this? What key elements are involved? How do they all fit together? How do we approach the solution? And most importantly, I got to drill down into the detail of what this actually looks like in practical terms. So much of the consulting industry is about high-level concepts, which are sound in logic but designed and delivered by people who don't understand the detail of the practical complexities. As the saying goes, 'Nothing is impossible for the person who doesn't have to do it'.

Through this journey, I discovered that the solution was all centred around integration, and I began to broaden my thinking well beyond just that of an Operations Centre. It became clear that what was required was a fully integrated operating model, and I focussed my attention on defining that. However, the closer I got to finalising exactly what this looks like, the more I came to realise it was worthless unless you could shift the mindset of the leadership team and take them on the necessary journey of transformational change. So, at this point, in 2015, after five years as an independent consultant, I partnered with the best transformational change consultants I had personally ever come across, to tap into their decades of experience in the area of leadership-led, codesigned transformation. And we co-founded NextGenOpX (which stands for 'next generation operating platform'), a management consulting company with a difference.

Combining our uniquely holistic, integrated operating model with the leadership-led transformational change capability, and backing this with a group of deeply experienced operational thought leaders, it has proven to be a knockout combination, one that is unmatched anywhere globally. We have subsequently demonstrated our ability to successfully deliver large projects on a global scale with major mining clients, including:

- developing a 'future operating platform' for various large mining companies;
- managing the global integration of companies following an acquisition/merger;
- and implementing a fully integrated operating model across multi-mine operations.

The depth of experience and detail developed as part of these and many other case studies has allowed us to develop a comprehensive understanding and methodology for designing and implementing fully integrated operations. In other words, there is real substance behind everything written in this book.

What drives and motivates us is the need (and opportunity) to change the industry for the better.

The book is based on this deep operational experience across the mining industry. It is based on observations of what works well and what doesn't; what companies do well and what they don't; what has changed and why; and most importantly, deep insights on what could and should be done better and how. I have been at the forefront of the subject of integration for over fifteen years now and have dedicated this time to developing deeply thought-through solutions, not just vague concepts.

This book is aimed at the mining industry and largely uses mining examples, because that is where my experience has been. However, the principles of integration are generic and have proven to apply in many industries and levels of complexity. You can apply them to the way a global corporation is set up, or to a single operation, a department, or a support function, whether it is in mining, oil and gas, utilities, or a service-provider business. The principles of integration should form the basis for ensuring alignment, integration, and excellence within every business. Although the principles were developed and refined empirically through a decade of multiple, diverse applications, they align well with most management philosophies, but also introduce a significant number of new insights and design principles.

At times I present quite a critical view of the mining industry. Some comments may come across as harsh, and some points are stated bluntly for effect. But the opinions I express are based on thousands of observations and hundreds of conversations with employees and stakeholders, at various hierarchical levels, across dozens of companies, all around the world, over four decades. As such, I am confident that the essence of the observations and comments are a valid assessment, even in the most advanced companies. There are certainly exceptions, but where excellence exists, it exists in pockets, and I have not found a single mining company that can be regarded as truly excellent with respect to fully integrated operations. However, a few have had the courage to start this journey after seeing the inevitable benefits of doing so, and they are reaping the rewards.

There are many good and admirable things happening across the mining industry. The purpose of this book is not to give a balanced assessment of the industry generally. It is to highlight some glaring problems that are not being addressed. And all of these centre around a lack of integration.

The first step in embracing transformational change is to con-front the problem and recognise the urgency. So, let's move to the first chapter, 'The Industry Has a Serious Problem and Is in a Vulnerable Position'.

CHAPTER 1 //

THE INDUSTRY HAS A SERIOUS PROBLEM AND IS IN A VULNERABLE POSITION

//

1.1 Productivity within the Industry Is Declining

1.2 Operations Are Not Getting the Basics Right

1.3 Technology Is Growing Exponentially Fast

1.4 There Is a Big Gap Emerging That Is Not Being Addressed

1.5 Businesses That Don't Adapt Face the Real Risk of Disruption

1.1 PRODUCTIVITY WITHIN THE INDUSTRY IS DECLINING

OBSERVATIONS

The mining industry has a serious problem with declining productivity and a lack of sustainability in its improvement efforts. The importance of this is masked when margins are high during high commodity prices; however, the cyclical nature of these prices presents a real risk to companies that don't address efficiency and productivity.

Various analytical studies have been carried out on mining industry productivity over the years, which all say basically the same thing: there was a steady improvement up until the early 2000s, and then a significant net decline over the following two decades.

Analysing productivity across an industry in absolute terms is a challenging process, because it depends on a lot of assumptions. The more detail you go into, the more questions you raise, and the more opportunity for errors, challenges to the assumptions, and confusion. For example, some of the productivity decline has been disguised by economies of scale and changes in grade. It's much safer to look at specific relativities and the level of change from one period to another. This cuts through many of the complexities of the analysis and provides a much stronger basis from which to draw conclusions.

The consulting companies doing these analyses have approached it from different angles, including from an equipment, labour, and capital perspective. Whilst some of the details have differed, the general conclusion is consistent in relative terms. Productiv-

The Mining industry follows a very reactive approach to managing through the price cycles

- What worked for the industry before, isn't working now
- Productivity has declined, despite advances in equipment design, higher technology, automation, data analytics
- The industry's reactive approach creates huge turmoil in businesses, and destroys value
- There is a greater need than ever now, for a more systemic, sustainable approach to managing operations

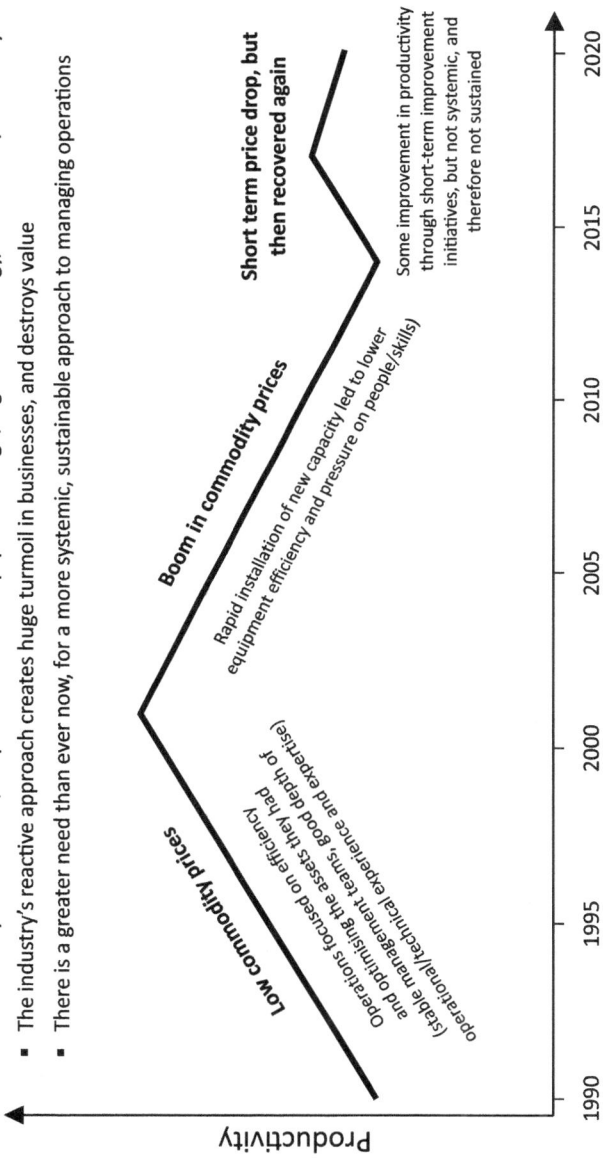

Low commodity prices

Operations focused on efficiency and optimising the assets they had (stable management teams, good depth of operational/mechanical experience and expertise)

Boom in commodity prices

Rapid installation of new capacity led to lower equipment efficiency and pressure on people/skills)

Short term price drop, but then recovered again

Some improvement in productivity through short-term improvement initiatives, but not systemic, and therefore not sustained

Productivity (y-axis)

1990 · 1995 · 2000 · 2005 · 2010 · 2015 · 2020 (x-axis)

Fig 1.1 Productivity within the industry is declining

ity increased up to and through the '90s, when commodity prices were relatively low and operations had to focus on efficiency and optimising the assets they had. Management teams were stable, and there was good depth of experience and expertise in technical and operational teams.

Then came a sustained boom in commodity prices in the early 2000s. Access to capital improved and mines acquired new capacity to address the urgent demand for growth. In many cases, they overcapitalised and bought equipment they weren't able to use effectively. Increasing equipment fleets in an open pit or underground mining operation increases the density of equipment activity in a confined space. This can quickly become a constraint where physical and organisational interactions start to cause conflicts if the designs are not fully thought through. At the same time, demand for experienced people increased rapidly, which led to much greater movement of people between operations and companies. This had an impact on the depth and tenure of expertise at mines, and it began to affect the fabric and quality of decision-making within management teams.

The net result of these two factors was that productivity levels dropped steadily through the 2000s and into the early 2010s. There was some short-term kickback around the mid-2010s when commodity prices dropped again for a while, and people started to use the newly installed capacity better. But then when prices recovered a few years later, productivity levels dropped again.

You could argue that this is just a logical reaction to price changes. However, it does beg the question: Do mining companies have to be as reactive as they are? Couldn't they do a better job of preparing for such swings? Everyone knows the swings occur in some type of cyclical pattern. Is it really necessary for large, sophisticated companies to react in complete surprise every time it happens and put their entire organisations into crisis-management mode?

These are things that should receive more attention from senior leaders. It isn't unreasonable to think that organisations can do a better job of managing through these cycles. It would help to avoid the huge turmoil and confusion, which is currently created every time a market downturn or upturn occurs.

There is no reason why the principles of efficiency, productivity, operational excellence, operating systems, management pro-cesses, and organisational models need to be turned upside down every time there is a market change. Yes, the way they are applied might change depending on the business priorities, but to turn everything upside down every time is just madness.

It highlights a fundamental weakness in operations at the moment. Whilst businesses and operations abound with a mul-titude of improvement initiatives, the industry as a whole has yet to crack the conundrum of continuous, sustainable improvement. The fact is, productivity remains a big challenge, even in the face of many other advancements, including larger-scale equipment, higher technology, more automation, and improved data ana-lytics. Most other industries have not shown the same volatility.

The mining industry continues to follow an approach that worked well a few decades ago, but it is not working as well now. It needs to set itself up to be more proactive and systemic in its approach and less reactive. Actions like arbitrary cost and head-count cutting, high-grading the ore body, and deferment of critical maintenance and sustaining capital are examples of reactive behaviour that produces short-term benefits but long-term value destruction.

The lack of continuous, sustainable improvement in productivity presents a significant risk to many companies when commodity prices go into a downturn, and equally, it presents a great opportunity for those who can crack the problem. The prize is enormous, both in absolute financial terms for every company and in relative terms through the opportunity to differentiate from competitors.

The industry needs to break away from its focus on short-term thinking (i.e. cutting everything back when times are tough, and pushing everything to the max when times are good). This boom-or-bust philosophy might make short-term sense, but it comes at a price, and there are many longer-term impacts that make it a losing game in the long run. Performance, costs, recruitment, retention, safety, risk, and governance are all affected.

There is a better way of setting up operations than the way they are now.

PRINCIPLES

Productivity is defined as the effectiveness of productive effort, or the rate of output per unit of input.

There are several ways to improve productivity.

Technical-related:

1. **Upgrading equipment through larger capacity units:** This is widely used and very effective, but often not done with the whole operating system in mind. Therefore, these units can result in increased productivity of individual equipment but lead to reduced overall system flexibility in dealing with certain situations. For example, where there is mismatching of the capacity of trucks and shovels, or shovel size/flexibility and ore-blending requirements.

2. **Improving technology through automation:** Autonomous equipment is increasingly used and is effective in increasing production rates if approached correctly. Ironically, the people-impact is often overlooked, which leads to the full value not being leveraged. So, a technological system needs to be approached with care and with a whole-system perspective, not just as a simple technology fix.

3. **Improving technology through algorithmic control:** Advanced process-control systems are the ultimate in productivity optimisation, because they are automated and can be programmed to optimally deal with multiple, complex scenarios. This is effective where the whole operating system is fully digitised, for example in processing plants. However, this is not currently the case in most mining operations, due to the current fragmented, batch-style nature of the mining process. As autonomous equipment becomes more deeply embedded and integrated into operations, this will become an area of great optimisation potential.

People-related:

4. **Pushing harder:** The 'big stick, loud voice' approach is the traditional style in many operations. This can be very effective, as long as you have such leaders, and as long as your employees are prepared to tolerate such a style. However, this is less and less the case in today's world.

5. **Improving motivation through incentives (rewards and bonuses):** This can work for a while, until they become expected; then they lose their impact. They can also be counterproductive if used inconsistently and ineffectively, because they often end up improving one metric at the expense of another. Such systems are usually set up with good intentions, but poor design and/or implementation ends up driving the wrong behaviours.

6. **Making it easier for people to work efficiently and effectively:** Making sure people are informed and aligned around a common goal is a massive area of untapped opportunity. The industry has barely looked at this to date, and they need to put a lot more effort into improving this area.

This book addresses all of these productivity principles in one way or another. Whilst the most obvious connection is to point 6, it also helps you to manage points 4 and 5 sensibly. And the connection to new and upgraded technologies (points 1 to 3) is also very strong, as it helps position operations for the more sophisticated working environment that will be coming in the future.

The key points are that the industry is not doing well with respect to productivity, and there is a big area of untapped opportunity associated with making it easier for people to succeed through better informing, aligning, and inspiring their efforts. Creating a working environment that pulls everyone towards new technologies will help to address one of the big weaknesses of current efforts in this space, i.e. the people-side, and change management.

SOLUTION

It would appear to be a no-brainer to align everyone and everything in the business around delivery of a clear business/operating strategy, and clear all other obstacles out of the way. However, most businesses do anything but that. In fact, most organisations are designed in a way that drives siloed behaviour and pulls everyone in different directions.

Why wouldn't you address that? Over the coming chapters, this book systematically explains the detail around what is missing, and how to go about setting up your business for greater sustainable success.

CHECKLIST

- How do you measure productivity in your organisation?
- What are your current productivity levels and recent trends?
- How do these compare with twenty years ago?
- Do you understand the reason for the differences?
- Are your current business improvement efforts aimed at achieving specific productivity targets?
- Are these targets really meaningful in terms of business value?

1.2 OPERATIONS ARE NOT GETTING THE BASICS RIGHT

OBSERVATIONS

The industry is generally not good at many fundamentals of operational excellence. Both productivity trends and anecdotal evidence indicate this is generally getting worse, not better. Improving the following basics will be key to the success of any transition to a higher-technology future.

Integrated planning: Planning is typically carried out in a relatively siloed way and then patched together in a coordination process. There is naturally some interaction between the different parts, but often time pressures result in autocratic decisions to resolve conflicts rather than dealing with the core issues behind the conflicts. Where integrated planning is done well, it brings together different functions (e.g. development, production, logistics, and maintenance). However, this is usually around specific bottleneck areas of the operation rather than the whole end-to-end operating system. And integration is usually limited to the value chain rather than extended to include other numerous aspects requiring integration, including, for example, between the long-, medium-, and short-term time frames.

Operating discipline: Plans are commonly viewed as guidelines by the frontline supervisors and operators, developed by 'out-of-touch' planners and schedulers who 'don't understand the real world' at the front line. There are many degrees of freedom in how things are implemented each day, and these create misalignment between the different parties involved (e.g. between production and maintenance). Lack of compliance to plans cre-

The mining industry is generally not good at many of the fundamentals of good operations

- **Integrated planning**
  - Typically, relatively siloed and poorly integrated, with long and short-term disconnects, and dynamic changes poorly managed
  - Instead of optimised, integrated, dynamically managed and realistic

- **Operating discipline**
  - Typically, fire-fighting, short term priorities win, poor conformance to plan, dynamic decisions siloed
  - Instead of disciplined execution, with dynamic changes addressed in a whole-system-value context

- **Stable, predictable operations**
  - Peak local performance is rewarded, instead of steady, optimised performance of the operating system as a whole
  - The true impact of variability and lack of predictability, is generally poorly understood

- **Standardisation, sharing, replicating**
  - There is relatively little standardisation across operations, even within a single business unit; limited learnings and value-transfer
  - Standardisation is generally desired, but is not supported by the necessary systems and processes to enable it to happen

- **Emphasis on the short term urgent over the long term important**
  - Short-term results tend to dominate the conversation, with multiple initiatives, addressing the symptoms not the cause
  - The mantra of 'long-term shareholder value' seems to have become replaced by 'short-term share price'

- **Continuous sustainable improvement**
  - Typically, improvement is viewed as very secondary to planning/execution, focussed on short-term results, accountability weak
  - Instead of a focus on continuous, systemic, sustainable improvement

Fig 1.2 Operations are not getting the basics right

ates a vicious circle back to the need for constant rescheduling and less time for optimisation, which leads back to less realistic schedules and plans, and so on.

Stable, predictable operations: One's life in operations can often feel like a roller-coaster ride, where everything is going well until some unfortunate and unforeseen event occurs—usually a breakdown caused by pushing the equipment or system too hard. Despite the fact that this happens constantly, we never quite accept that this is the norm. We somehow convince ourselves that the event is some sort of aberration. It isn't. It is a natural outcome of prioritising peak performance of the parts of the operating system over steady and optimised performance of the operating system as a whole. The reason this occurs is because we encourage it. Operators are rewarded for breaking records, even if it is at the expense of the next shift, next day, or next week. And so, another vicious circle is created.

Standardisation, sharing, replicating: Standardisation makes obvious sense. However, the reality is that there is very little of it across the industry, even within companies across different operations. Even between shifts in the same operation! Everyone loves the idea of standardisation, 'as long as you standardise it my way, because I've given it a lot of thought'! It's the classic 'not invented here' attitude. There are real behavioural obstacles that stand in the way of standardisation, sharing of ideas, and replication of improvement, but there is little focus on these. It's important to remember that standardisation is a foundation for the application of technology.

Emphasis on the short-term urgent over the long-term

important: Short-term results tend to dominate the conversation, even when they occur at the expense of long-term results. The list of improvement initiatives usually focuses on point solutions, not systemic solutions, which address the underlying issues. This focus on the short-term starts from the very top of the organisation. As previously mentioned, the mantra of great companies used to be 'long-term shareholder value', but the focus of most businesses has unfortunately shifted to a narrower focus on short-term results. This has been driven by more reactive shareholder demands and news cycles.

Continuous sustainable improvement: Underpinning all of these observations is the fact that operations struggle with continuous, long-term, systemic, sustainable improvement. And, without this, it is difficult to break out of the cycle of inconsistent performance and continually reinventing the wheel.

PRINCIPLES

The problem with not getting these fundamentals right is that they are the foundation of good operations:

- **Integrated planning:** If you don't plan in an integrated way, how can you align all the different parts and optimise the value of the operating system as a whole?
- **Operating discipline:** If you don't execute the plan in a disciplined way, it forces you into a vicious circle of constant rescheduling with no time to optimise.
- **Stable, predictable operations:** If your operations are not stable or predictable, you don't have a base on which to optimise and improve.

- **Standardisation, sharing, replicating:** If you don't standardise your operating practices and share improvements, how do you avoid constant reinventing of the wheel? Note that just creating standards doesn't necessarily lead to standardisation; there are social processes involved in the latter.
- **Emphasis on the short-term urgent over the long-term important:** Don't expect long-term improvement if you just focus on short-term results.
- **Continuous, sustainable improvement:** Systemic improvement requires a different approach, which addresses the *cause* of problems rather than the *symptoms*.

These principles are covered in detail in Chapter 3.

SOLUTION

The answer to these problems is to approach these key fundamentals in a sensible way.

- Focus on optimising the system as a whole, not just pushing the parts to the limit.
- Ensure your planning is integrated across the whole operating system, is optimised for whole-system value, and is realistic.
- Execute the plan in a disciplined way with an emphasis on compliance to plan.
- Create a single, integrated schedule that covers all key activities, and manage this dynamically (i.e. in a way that accommodates the constant unforeseen changes, which invariably happen in normal operations).
- Reward performance of the operating system as a whole, not individual performance.

- Be clear on the best way of operating each part of the system and the whole, and design your management processes to encourage a consistent and standardised approach.
- Focus as much attention on systemic improvement as you do on short-term results.
- Don't just rely on rhetoric. Systemically embed all of the above into the management processes, systems, and organisational architecture of the business, so this is constantly driven and becomes established as the culture.

CHECKLIST

- Do you specifically plan for performance of the whole system, or do you just aggregate its parts?
- Is this an ongoing integration/reconciliation process, or just an annual planning requirement?
- Do you measure compliance to your plans and schedules?
- Is compliance to plan a primary driver, or is it secondary to short-term results?
- Do you measure variability in performance of the whole operating system and its parts?
- Do you measure variability in performance between shifts (as a measure of standardisation)?
- Do you track performance improvement trends over multiple years or just year-on-year?

1.3 TECHNOLOGY IS GROWING EXPONENTIALLY FAST

OBSERVATIONS

New technologies and their impact on businesses and societies are developing at an incredible pace.

The technologies themselves cover a wide range of areas:

- Automation (autonomous equipment, automated processes, robotics, drones, etc.)
- Sensors (smart sensors, 'Internet of Things', etc.)
- Digital platforms (end-to-end business systems, interactive connectivity, data integration, etc.)
- Data analytics ('Big Data', simulation modelling, predictive analytics, etc.)
- Artificial intelligence (Quantum computing, machine learning, etc.)

These technological advancements are real. They will affect every operation, and they are evolving fast. You had better be prepared for this, because they will certainly affect your business in one way or another.

However, the social impacts of these and other technologies rapidly flooding our society are equally significant:

- Social media and smartphone communications
- Remote work and flexible working arrangements
- Changing expectations of the younger generations
- Changing demographics in the workplace
- Regional socioeconomic challenges, environmental/ethical factors and geopolitics

There is a plethora of new technologies becoming available to the Mining industry, along with with major social shifts and changes in management culture

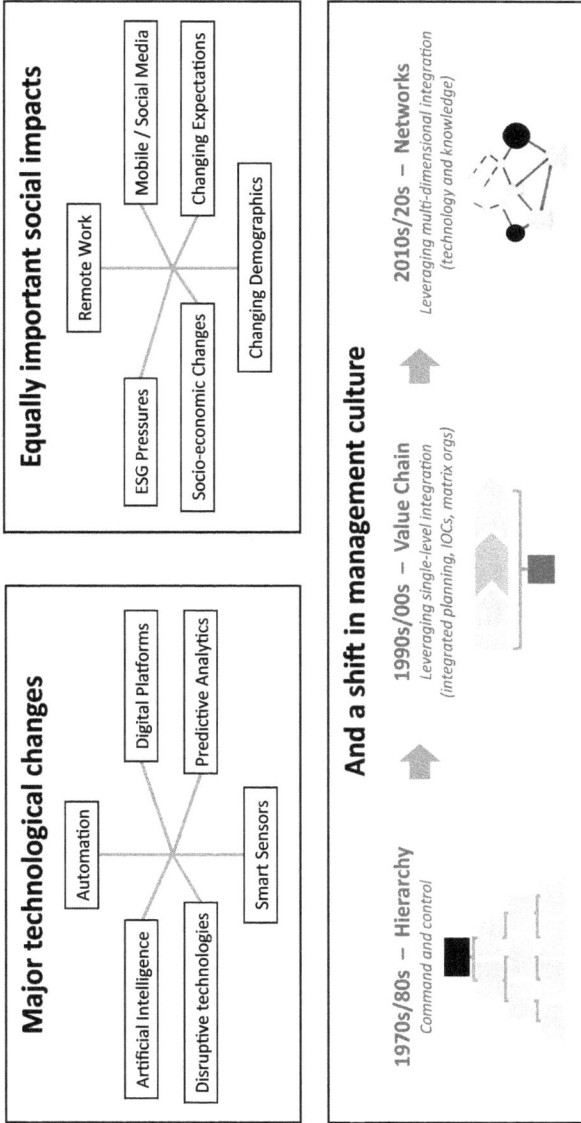

Equally important social impacts

- Remote Work
- Mobile / Social Media
- Changing Expectations
- ESG Pressures
- Socio-economic Changes
- Changing Demographics

Major technological changes

- Automation
- Digital Platforms
- Predictive Analytics
- Artificial Intelligence
- Disruptive technologies
- Smart Sensors

And a shift in management culture

1970s/80s — Hierarchy
Command and control

1990s/00s — Value Chain
Leveraging single-level integration (integrated planning, IOCs, matrix orgs)

2010s/20s — Networks
Leveraging multi-dimensional integration (technology and knowledge)

Combined, these add up to a profound shift in the way operations need to be managed

Fig 1.3 Technology is growing exponentially fast

- Growing ESG pressures (environmental, social, and governance)

The cultural style within operations has also been steadily shifting over the past few decades:

- From hierarchical (command and control) in the 1970s and 1980s
- To a value chain focus (leveraging technology and knowledge) in the 1990s and 2000s
- To connected networks (leveraging rapid decision-making and collaboration) in the 2010s and 2020s

All of these technological changes and their social impacts add up to a huge shift in the way operations need to be managed in the industry. The pace of these changes will likely increase, not decrease, in the future.

The next generation of leaders will have different expectations of their operating models. They will demand speed, connectivity, and seamless design. This is a long way from what currently exists in most operations.

Building a new technologically and socially advanced digital platform, and building a new operating model, must happen concurrently. The problem is that one relies on the other. You can't fully transform a business without considering both together. The end result is complicated enough to design, but managing the transition from the old to the new adds significantly to the complexity and risk. This is currently a major gap in the technology/digital strategy of most businesses.

The bottom line is that the rapid introduction of new technologies presents both an opportunity and a risk. It will involve a lot more than just the technology to leverage the full value from any changes.

PRINCIPLES

The rapid approach of new technologies is a certainty, and operations ignore this at their peril.

However, there is no doubt that much of the current commentary in this space includes a lot of 'hype', with little appreciation for what exactly this means for operations in practical terms. The reason this space gets such traction is because it is easy to get excited about technology and the potential benefits of a fully automated process. This is true particularly for the person who is pushing the technology rather than the one having to manage the change. Great technologies often fail or don't leverage their full potential value due to poor change management during both the design and implementation stages.

'Automation' and 'digital' are being presented too often as a magic solution to the people-challenges within operations. They are not. 'Automating' a poor operating system is simply going to automate that poor operating model.

Frankly, most operations are currently nowhere near stable and predictable enough to deliver a transformational change in this regard. Autonomous truck and drill fleets, yes, but fully connected and automated mining systems, not yet. Digitised planning and remote Integrated Operations Centres, yes, but fully controlled decision-making from the centre, not yet.

Even in a greenfield application (i.e. a completely new operation, which can be designed afresh), designing a fully connected, automated, digitised, and algorithmically controlled operation is, at the moment, a bridge too far, and there are none yet who have come even close to this. But in a brownfield application (i.e. an existing operation, which needs to change from its current approach), trying to retrofit a completely automated system would present another level of challenge altogether.

Businesses targeting true transformational change in this way will need to focus more on fixing the operational weaknesses that need to be addressed, in order to underpin any progressive technology strategy.

SOLUTION

The key to getting your technology/digital strategy right is to understand that it is a journey, not a destination. The most common mistake businesses make is to just have a long list of technology projects that are not integrated and do not consider the implementation and change management challenges.

Another common failing is to focus more on the specific technologies than on the underpinning operating platform. Millions of dollars are spent on testing the technologies themselves, but rarely is much money spent on improving the working environment in which they will sit. Mining businesses are quick to push new systems, processes, and technologies onto operations to use, which are user-unfriendly, unintuitive, and don't interact well with other parts of the operating system.

Given that the biggest challenge of new technologies is often not the technology itself, but the adoption of it by operations, it is important for technology and digital staff to pay much greater attention to this aspect. Empathy for the operational realities is important, while not allowing these to prevent you from breaking out of a paradigm way of thinking.

Lastly, an advanced technology programme, which is built on an unstable or unpredictable operating platform, is never going to be successful. It is important to get the operating basics right first (or in parallel), and make sure your operation is in control. These nontechnological basics are often the hardest problem of all.

CHECKLIST

- Is your technology and/or digital strategy more than just a list of technology and/or digital projects?
- Have they been crafted into an overarching, staged, short-, medium-, and long-term roadmap that considers the evolutionary journey, including the operationalising of the technologies and supporting operating model?
- How well are these strategies understood, codesigned, and supported by the operations teams?
- How well are the technologies you are chasing typically adopted by operations (your own and others)?
- How well are the people and change management side of new technologies managed by operations (your own and others)?

1.4 THERE IS A BIG GAP EMERGING
THAT IS NOT BEING ADDRESSED

OBSERVATIONS

The big conundrum for the mining industry is the growing gap between the exponential rate of increase in new technologies and the steady decline in operational capability.

As already mentioned, technology is not a silver-bullet solution, which will replace the need for solid operating fundamentals. On the contrary, technology needs a solid platform on which to do its magic. The problem is that there is a dichotomy at the moment between these two spaces. Operations people play in their space and the Technology people play in theirs.

The Operations camp thinks the Technology vision is crazy because, 'We can't even get the operational basics right, so how can we expect to pull off a high-tech world?' And 'So many of these technology projects we have tried in the past have been fraught with issues, and these end up getting passed on to us in Operations to solve or live with'.

The Technology camp thinks the Operations mindset is crazy because, 'Technology is clearly the future, and it will help fix so many of the operational problems of today, so it makes no sense to resist it'.

This characterisation is overemphasised to make a point, but in essence it is true. This is not to say that there isn't good intention on the part of both the Operations and Technology camps.

Advances in new technologies are currently running way ahead of our ability to adopt them

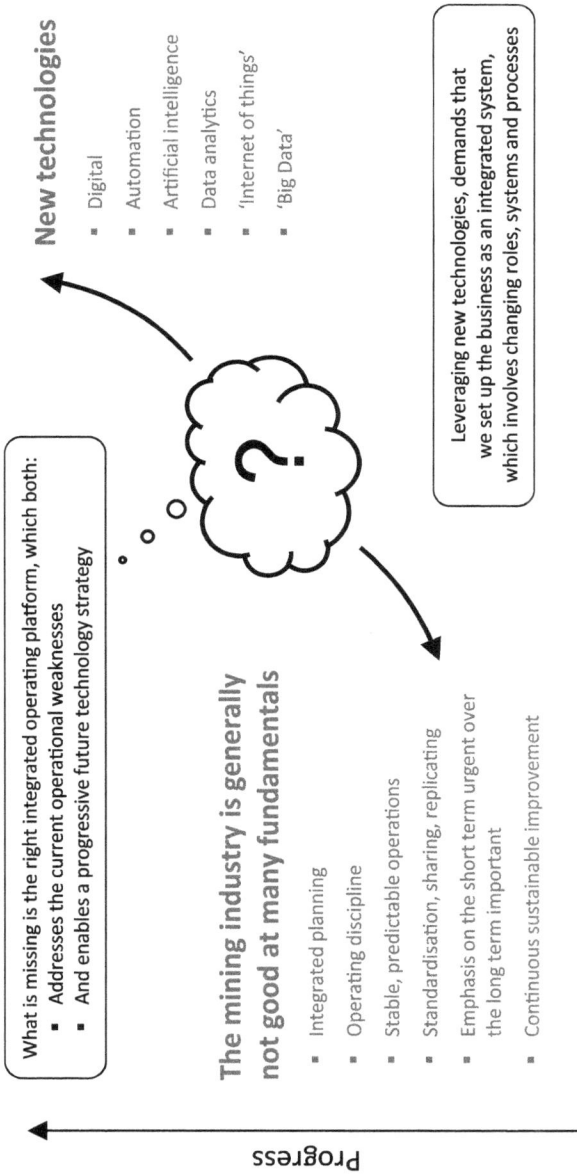

New technologies

- Digital
- Automation
- Artificial intelligence
- Data analytics
- 'Internet of things'
- 'Big Data'

What is missing is the right integrated operating platform, which both:

- Addresses the current operational weaknesses
- And enables a progressive future technology strategy

The mining industry is generally not good at many fundamentals

- Integrated planning
- Operating discipline
- Stable, predictable operations
- Standardisation, sharing, replicating
- Emphasis on the short term urgent over the long term important
- Continuous sustainable improvement

Leveraging new technologies, demands that we set up the business as an integrated system, which involves changing roles, systems and processes

Progress

Technology is not a silver bullet that will replace the need for solid operating fundamentals

Fig 1.4 There is a big gap emerging that is not being addressed

There are many on both sides who understand the challenges of bridging the gap. However, what is missing is the vehicle to do so.

Consultants and technology companies typically amplify the problem because they are fixated on the technology and the money they can make out of it. However, they bring few answers to how to address the fundamental operational issues. They sound flexible by saying, 'That is your turf; we don't want to prescribe; we can accommodate whatever you want in that space', but in reality, that is the hardest part of all. If it were that easy, Operations would have fixed it by now, and the problem wouldn't exist. And so, the gap remains. This is an area of major differentiation in the approach we adopt at NextGenOpX. We go after those areas, because we know this is where the heart of the challenge lies, and where simple, smart solutions to age-old problems are sorely needed.

What is needed is the right integrated operating platform that addresses the problems of today, and simultaneously positions for the future. We need one that brings clarity, simplification, focus, and discipline to the operating environment, and also one that provides a process for managing the transformational change required.

This missing link is integration.

PRINCIPLES

The concept that you need a stable operating platform on which to overlay technology is a completely logical one. Integration is the means to achieve that stability, and an integrated operating platform is the end result.

How can you expect to leverage the full value of technology when teams work in silos, planning is not fully integrated, compliance to plans and schedules is not disciplined, operations are not stable or predictable, and operating procedures are not standardised?

Once your process is fully continuous, automated, and algorithmically controlled, like in a processing plant, then a lot of the interfaces between people and processes go away. However, as long as you have a largely batch-type process, which is the current norm in most nonautomated mining operations, the way these interfaces are managed is absolutely critical to success.

This doesn't even touch on the massive difference in operating culture between a low-tech and high-tech operation. In greenfield projects, it is common to hear, 'We are recruiting a fresh team with a new mindset, who will embrace the next generation of thinking'. This is great in theory, but the reality is that these people are still being recruited from the mining industry labour pool, which is very much in a low-tech paradigm, especially from an operating model and operating culture perspective. And they are often being led by a person with a lifetime of experience in the 'old ways' of the industry. So, it is much easier said than done to 'start off with a completely new mindset'.

Nor does this touch on the challenge of transitioning a team from a low-tech to a high-tech operation. This is the biggest challenge of all for brownfield operations.

The significance of this gap between implementing technologies and operational adoption is well-illustrated by two examples.

- First, in a mining business where autonomous trucks were implemented at three different mines, there was a large difference in the productivity improvement achieved (30 percent, 10 percent, and 1 percent) as a direct result of the way the implementation was approached with the operations teams. The most successful applications were the result of the GM driving the change holistically and wholeheartedly. The less successful applications were a result of hesitant change management and drive.
- Second, in the area of mining software, there are countless outstanding systems available that are well-designed and have extremely powerful capabilities. However, it is a consistent comment from these software providers that their systems don't begin to add the value they could, because they are being used within a dysfunctional client operating model. These service providers reluctantly resort to focusing on 'implementation' of their systems, knowing that 'adoption' of their systems within the operation will be weak. Many of these companies are expressing great interest in the integrated operating model approach, because it improves the amount of value that can be leveraged from their systems and equipment.

So, finding a way to bridge this growing gap between the exponential rate of increase in new technologies and the challenges of declining operational capability is absolutely essential.

SOLUTION

What is missing is integration.

Integration between the parts of the mining process; between the planning, execution, and improvement functions; between the technology projects themselves; between the technology projects and operations; between the various stages in the journey from the current to future state; and between all the people involved in each of these areas.

It is not surprising that this doesn't happen naturally, even for an enlightened leader who instinctively aspires to achieve this. Without the benefit of a structured way of approaching this, they quickly realise that it is very hard to do, and they are better off pulling back and just focussing on their own area of responsibility.

The common link that allows you to align and simplify all of this is the 'lens' of integration. It is a different way of looking at activities, and it is common to all situations. Because of this, it allows you to align the approach across all the activities described above.

CHECKLIST

- How well-aligned are your operational teams with your technology and digital strategy?
- Are the technology projects being done *with* them or *to* them?
- Is there as much attention being applied to the journey from the current to future state as there is to managing today's operations and the technology programme itself?
- Does your business have (or need) a chief transformation

officer (or similar equivalent) to help navigate this complex journey?

- If you have a Transformation role, is it being led by one of your most capable/senior people with experience in both operations and technology and with a particular strength in people and communications?

1.5 BUSINESSES THAT DON'T ADAPT FACE THE REAL RISK OF DISRUPTION

OBSERVATIONS

So what? There is a gap. Why is that such a big deal? After all, the whole industry is in the same boat, so relatively speaking, there isn't a big risk to companies. A missed opportunity to lead the industry maybe, but no big downside. Right?

Wrong. This is a very big deal for the mining industry and any company that aspires to excel within it.

First of all, mining companies are typically leaving 20–40 percent value on the table due to the lack of integration and decline in their operating performance (not getting the basics right, declining productivity, and challenges with skills and experience).

But, more significant than this, the industry is becoming increasingly vulnerable to disruption. The gap between technology and operations cannot continue unaddressed. The opportunities presented by new technologies, and particularly complete digitisation of the mining industry, are enormous. So big, in fact, that someone will step in and seize the opportunity sooner or later.

Given the pace at which emerging new disruptive technologies are being developed, this is likely to occur within the next ten years or so. And there is a very good chance that this could happen from outside the industry.

Full digitisation, including a high degree of automation, would look something like what Amazon has done to the retail industry.

Turning the value chain into a seamless process, from raw materials (retail products), to supply chain management, to logistics, to the customer, all in a highly efficient, transparent, and mostly automated, user-friendly process.

Compounding this (and likely catalysing), are two potentially game-changing global technology disruptions that are very likely to disrupt everyday life (and mining businesses) in a major way over the next decade or so. These are the disruption of energy and transportation. In particular, the convergence of a number of technologies related to these two areas (batteries, electric vehicles, autonomous vehicles, transport as an on-demand service, and solar energy).

There is plenty of information available on these trends, backed up by well-articulated arguments with facts and figures, extrapolating just how disruptive this could be. It is worth researching the views of some of the leading commentators in this space, and assessing the potential implications for your business/operation. You may not agree entirely with the projected timelines for some of the predictions (despite evidence of the pace of previous technology disruptions being wildly underestimated), but the inevitability of the direction of this combination of technologies is clear. And to imagine this technological transformation taking place within the next ten years is not difficult.

What this will mean for the mining industry is a radically different way of setting up and operating mines and support centres. This will come about through fundamental changes to the cost base, particularly as it relates to transportation and the supply chain. Autonomous trucks and drills and even face-cutting machines

Change is coming (fast) – those that prepare well for it, will have a big advantage

The need to change is 3-fold:

1. Improving your current business performance (generating immediate short-term upside value)

2. Managing your transition to the high-technology future (avoiding medium-term downside value)

3. Positioning yourself with strong options, when disruptive change occurs (major long-term competitive advantage)

You have 3 options:

- Prepare now for the inevitable – **LEADER** (capitalise on the full value opportunity, and enhance your reputation)

- Wait until you see it happen – **FOLLOWER** (miss value in the interim, and potentially disenfranchise your workforce)

- Wait until you are forced to change – **BYSTANDER** (most probably miss out altogether, and put your business at risk)

Tips for success:

- Appreciate that the world is changing at an incredible pace, and what works now will not always work

- Focus on getting your operating fundamentals right

- Ensure your digital/technology strategy is fit-for-purpose and something which your Operations teams buy into

- Build the right operating model, which drives short term value and positions you well for the technological future

- Use integration as the vehicle for becoming informed, stable, predictable, standardised, systemic and sustainable

- Recognise the problem and the opportunity; don't delay; get started on a holistic strategy right away

Fig 1.5 Businesses that don't adapt, face the real risk of disruption

(continuous mechanical mining equipment) are not new to mining, and there has been good progress made with these over the past ten to fifteen years. But what we are talking about with a global disruption of energy and transportation will occur at a much faster pace.

Mining companies are not used to moving at lightning speed and are, for the most part, not experts at supply chain logistics. This fact will not be lost on the technology companies that emerge as leaders in this space. They will see the opportunity to enter and disrupt the mining industry with these new technologies and methodologies. And there is a good chance that much of the industry could be left standing.

PRINCIPLES

Existing mining companies have, to date, shown little sign that they have the capability to truly transform the industry through full digitisation. They are so distracted by the huge challenges of running their operations efficiently in the way they always have, that they are missing a much bigger picture. This is likely to be the same for other industries as well.

One of the paradigms miners struggle to break free from is the mindset that mining is immensely complicated, with many unattached moving parts, and many degrees of freedom at every part of the operation. The reality is that mining involves analysing an ore body, working out the optimum way to mine it, and then mining it that way.

The industry actually does the first piece very well. The life-of-mine (LOM) planning is the only time when the operation is truly looked at from an integrated, whole-system perspective. Each block in the ore body is completely understood (from a geological, mineralogical, chemical, hardness, head grade, concentrate grade, recovery, impurity, etc. point of view), and then cost, efficiency, value, and risk factors are considered (like depth, distance, equipment, methods, and logistics), in order to arrive at the all-round optimum way of mining that ore body.

Unfortunately, that is where the optimisation ends. What happens after that is that disconnects occur during the planning and execution processes, due to too many degrees of freedom and insufficient governance. As such, the ore body often becomes suboptimally mined over time, and a vicious circle is created where siloed planning and execution decisions further erode the quality of the planning and execution. Then the whole operation starts to become fragmented, misaligned, unpredictable, and unstable. This is compounded by well-intentioned but poorly applied efforts from leaders to constantly squeeze out short-term results locally, even if it is at the expense of long-term value across the whole system. Because the business systems and processes themselves are relatively weak, it is not possible for the organisation to hold all of these permutations together dynamically. Stewardship of the resource suffers, and the result can be a bit of a mess.

What should happen is that after developing that fully optimised LOM plan, block-model, and block-processing schedule, the whole organisation should be turned into a 'sausage machine'. In other words, a highly predictable connected process that is totally focussed on processing those blocks in that order. Minimal degrees of freedom, a ruthless culture of compliance to the plan, absolute transparency of dynamic changes, and centralised control of whole-system value.

This is what companies like Amazon have done very successfully. They have far more moving parts than in mining, in an uncontrolled working environment (every street and home across the whole world!), yet they have turned logistics into a highly reliable 'sausage machine'. It doesn't take a wild imagination to recognise that companies like Amazon could do exactly the same in mining. Perhaps this is an oversimplistic comparative example, but there is no doubt that we currently overcomplicate our lives in mining, through our tolerance of too many degrees of freedom in the way we set up and manage our operations.

SOLUTION

The mining companies who position themselves the best for this future will be able to turn this threat into an opportunity. Positioning themselves will not be about trying to be advanced in new technologies, because they will never be able to outcompete global technology companies in that regard. Rather, it will be more about being the best at the basics of mining, and establishing an operating platform that is ready to move quickly with new technologies when they come. As an absolute starting point, this has to mean being great at the basics and having stable, predictable, and standardised operations.

This is an important distinction for mining businesses to consider when shaping their business and technology strategies. In all cases, a fully integrated operating model should be a fundamental requirement in this regard.

CHECKLIST

- Has your business considered the impact of this sort of technological disruption in its strategic planning and future scenario analysis?
- What would your business need to do differently to position itself to turn this from a threat into an opportunity?

STRUCTURAL WEAKNESSES NEED TO BE ADDRESSED

//

2.1 Complexity Is Increasing and Capability Is Decreasing

2.2 There Is More Business 'Clutter' and Less Clarity of Purpose

2.3 Behaviour Is More Siloed than Collaborative

2.4 Senior Leaders Now Focus Too Much on the Short Term

2.5 Current Business Improvement Models Are Not Working

2.1 COMPLEXITY IS INCREASING AND CAPABILITY IS DECREASING

OBSERVATIONS

Why is operational performance such an issue today? Why has productivity declined? Why can't we sustain improvements? Why is standardisation so difficult? Why do we have to keep talking about getting back to basics, and what does that even mean?

These things didn't seem to be such a challenge twenty to thirty years ago. What has changed? This is a very important point to understand before trying to address the questions.

Basically, the answer is that there has been a huge shift in complexity and capability over the past twenty years, which the industry has not recognised or reacted to.

By all accounts, these challenges apply to many industries, but let's explore more specifically what has happened in mining.

Complexity: Mining operations are substantially more complicated to manage today than twenty years ago. Mining operations are bigger, often run as multiple mines with blending requirements, more geographically spread, and increasingly in remote locations or high altitudes. Ore bodies are deeper, wetter, lower grade, with more impurities, and increasingly moving underground. Safety is a much bigger deal, as are health, the environment, and the community. Major project accountabilities are sometimes also added to the management workload. Bureaucracy, compliance requirements, and corporate initiatives have become more of a drain. ESG pressures and corporate gov-

Mining operations are substantially more complicated to manage today than 20 years ago

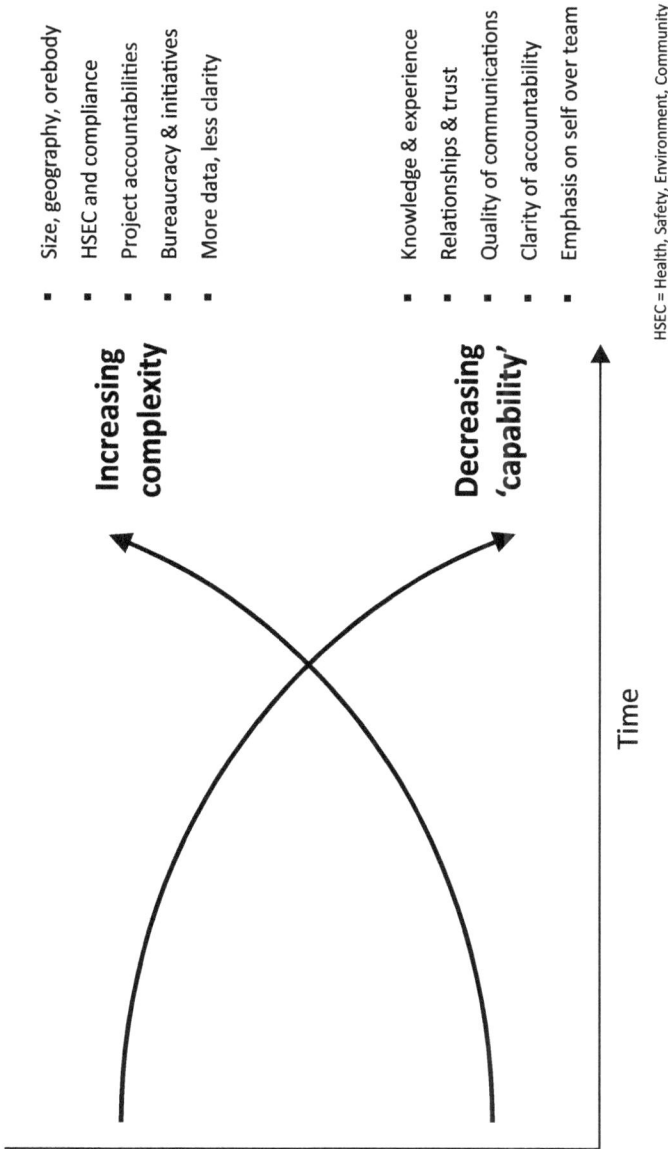

Increasing complexity

- Size, geography, orebody
- HSEC and compliance
- Project accountabilities
- Bureaucracy & initiatives
- More data, less clarity

Decreasing 'capability'

- Knowledge & experience
- Relationships & trust
- Quality of communications
- Clarity of accountability
- Emphasis on self over team

Time

HSEC = Health, Safety, Environment, Community

Fig 2.1 Complexity is increasing and capability is decreasing

ernance expectations are greater. The amount of data available is exponentially higher, yet somehow we still seem to have less clarity on how well our operations are running.

Capability: Capability has declined considerably over the same period. Not the inherent competence and capability of individuals, which has arguably increased, but the dynamics within an operation's management team. If you go around the table of an operation's management team today, you will often find a group with relatively short operational experience (due to more rapid promotions), and less tenure at that particular mine site (due to greater movement of leaders between roles and businesses), compared to much greater stability in the past. When people are only at an operation for a short while, it impacts the way they think and what others think about them. It has an effect on relationships, behaviour, communications, and trust.

Both of these issues (complexity and capability) are realities of life today. They are certainly worth being more aware of, but there is not really too much one can do about either in their own right. So, what is the point of raising it? The point is that the combination of the two represents an enormous shift in the challenges of running operations. Yet we are still running operations in the same old way we always have. And it's not working! Back to the underlying problem that this is not working out too well for most operations, in terms of productivity performance and operational excellence.

What has changed is the level of integration. When you have a team who have been together for a long time, who have deep experience, know the operation very well, and have a tight relationship with each other, then integration happens naturally. You know each other's problems, you communicate well and openly, and you're less inclined to do something in your area that will be positive for you but negative for another area. Plus, institutional knowledge is retained and recalled.

When you have a team of less experienced individuals who don't know and trust each other so well, this integration doesn't happen naturally. So, you need to compensate for this; otherwise performance will be affected. This shift in team makeup is exactly what has been happening throughout the industry over the past two or three decades, but there has been no adjustment for it, and the impact has been magnified by the increased complexity of operations.

The conclusion is that there is now a greater need than ever to invest effort into systematising everything you do. This means building greater structure into your systems, processes, accountabilities, planning, improvement, decision-making, and knowledge management. Again, the common link that binds this all together is integration.

PRINCIPLES

The life of a general manager (GM) of a mine is very different today, compared to a few decades ago.

In those days, it used to be a fairly autonomous role, left to its own devices, as long as the results were being delivered. Many of the complexities of today were not around to the same degree, and the management team had deep experience. The role of the mine GM was, to a large extent, an integration role. They spent a great deal of time aligning and integrating the efforts of the team and ensuring the whole operating system was run in a disciplined and optimised way.

Today, the role of mine GM is often so distracted by corporate issues and community responsibilities that they are barely on site for a lot of the time. Their team is comprised of less-experienced managers with individual career aspirations, who don't instinctively align well as a team. Add to this the further pressures of larger-scale roles and greater complexities, and the situation can quickly reach a breaking point. This is all compounded by the disruption created by higher turnover in roles across all levels.

And with the mine GM having less time to manage, resolve, and integrate all of this, the operating 'system' starts to break down.

The industry has, over a long period of time, rewarded managers who can come into an operation, quickly see the problems, focus on those problems, clear the decks of all other distractions, shuffle a few people, and drive the priorities hard with a strong voice of authority and a degree of fear. It works. It produces results. It is how managers have proven their value. And it results in a promotion to the next level.

Some experienced managers will say, 'That's all I need; I know how to get this done'. And they are right. Except for the fact that it isn't sustainable. They are proud of their ability to drive their team hard and get things done. But even they will admit that when they leave, it will be a different story. 'But that's not my problem', they say. First of all, managers who lead with a 'big stick and loud voice' are increasingly rare today, because this style is going out of favour. And secondly, it *is* their problem, or it should be.

The industry needs to shift to rewarding sustainability of performance—stable, predictable, and repeatable.

SOLUTION

The solution is to change the expectations of the leaders of today. Their roles should be to build systemic and structured resilience, and a culture of sustainable performance into their teams. They should be rewarded for performance of the whole 'system', not just their area.

How different would the performance of our operations be if each manager were rewarded for the performance of their area in the three years after they have left the role? Not just operational leaders, but all functions. Particularly the senior leadership roles. Perhaps not practical, but interesting to imagine and reflect on.

This involves a new definition of what a successful manager looks like. Currently, the top of the pile is the kick-ass manager who gets quick results. We all secretly want and value such people who can get things done and make problems go away. But in the process, the planning and systemic improvement functions are often subordinated and undervalued. This needs to change.

CHECKLIST

- What is the experience makeup of your own leadership team?

- Do you measure or track/trend this?
- What behaviour do you drive in your leaders?
- How do you describe your expectations of them (does it even mention systemic/sustainable improvement)?
- Does your reward system encourage short-term, unsustainable results?
- How do you encourage a greater focus on systemic improvement?
- How strong is your succession-planning process, not just from a pipeline perspective, but also transition/handover?

2.2 THERE IS MORE BUSINESS 'CLUTTER' AND LESS CLARITY OF PURPOSE

OBSERVATIONS

One of the frustrations of operations leaders is the lack of clarity, both with regard to the business as a whole and their role within it. There are two aspects to this: clarity of purpose, and unnecessary clutter and bureaucracy.

Clarity of purpose: Of course, roles are generally well-defined in terms of area of responsibility, span of control, decision rights, etc. And the senior executives will swear blind that the leaders in their team are crystal clear on what is expected of them. However, it is common to hear from people a single level down, at GM level, 'I wish my boss would just tell me what they want and then leave me to get on with it. It drives me crazy when the rules and priorities constantly change without context, and my hands are tied behind my back with regard to how I manage the details of my area. Then I get beaten up when short-term decisions come back to bite us.'

At the frontline level it is not uncommon to hear, 'I don't know. I've given up. I have pointed out some of the basics that are wrong and what we need to do to fix them, until I am blue in the face, but no one listens. I just sit back now and wait for the next instruction from above.'

Now, you can just look at these examples as poor leadership and not something that would happen in your own operation. But they are very common views, especially when considered in the context of the increase in complexity and decrease in capability discussed in the previous section. Other comments from the front line illustrate this even further: 'We have had three GMs in the past year'; 'I have been without a manager for the past six months'; 'My new production superintendent has never had an operational role before'.

This emphasises the importance of a greater reliance on well-structured, systemic, standardised processes, which can withstand regular leadership changes. The bottom line is, your systems and processes should not be dependent on the leaders. They should be the backbone of the business, which the leaders work to, not there to be changed or ignored at the whim of each new leader who arrives. This backbone is how the clarity of purpose and roles can be embedded into the business in a systemic and governed way.

Clutter and bureaucracy: There are a multitude of business systems on offer, and the technology boom has increased both the number of alternatives and the pace of improvements. As a general rule, mining businesses tend to view these systems as a noncore cost of doing business rather than value-adding tools. This is unfortunate, as it leads to unnecessarily onerous and cumbersome designs that are designed more to make life easier for the people who manage the systems than for the people who use and interface with them.

As a result, business systems are often bureaucratic and painful to use. Compliance to systems and processes is poor. And many 'work-arounds' exist.

This is compounded by the fact that there is a constant churn of initiatives, each coupled with its own systems and processes, which get added to the plate of the teams on the front line over

Much more can be done to align everyone's efforts and simplify the working environment

Moving from ...

Haphazard – a bunch of initiatives

- Poorly designed and executed systems, processes and improvement initiatives
- Fragmented, misaligned, overlapping, gaps
- Creating duplication, frustration, wasted resources, unrealised value

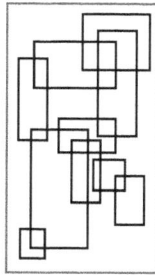

To ...

Well structured – MECE (Mutually Exclusive, Collectively Exhaustive)

- **Aligning everyone and everything in the business around the business strategy**
- **Removing the friction in the business:** Simplifying – Aligning – Integrating
- **Providing context:** Whole-system view – Transparency of info – Collaborative value improvement
- **Building capability and accountability:** Clarifying roles & expectations – Focussing – Disciplined execution

MECE is a way of organising things in as simple and clear a way as possible

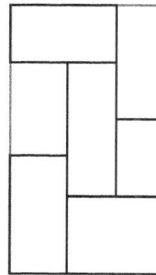

Fig 2.2 There is more business 'clutter' and less clarity of purpose

time. Seldom is there a clean-out of old systems, or a comprehensive review of what is important and what isn't. This all just adds to the burden and distraction for the operational teams.

In any well-run operation, clarity and simplicity are crucial for systemic, sustainable success. The easier we can make it for people to succeed, the more likely they are to do so. It is therefore worth focusing on this a lot more than we do at present.

PRINCIPLES

The key principle here is to do everything you can to make it easy for your teams to succeed. Make it clear and make it simple.

Make sure your teams understand what the overall operating system is—where it begins, where it ends, what all the parts are, how they all fit together, and what their role is in producing an optimised end result. Be clear on the rules of the game (whole system comes before area, before team, before self). Be clear on decision rights, mandates, and escalation processes. Be clear on what sort of changes you need to deal with dynamically and how conflicts are handled.

Primacy of the whole system is key. Build your systems around management of the whole system and the interfaces, not just its parts. Value the simplification of your business systems and processes, and user-friendliness. By user-friendliness, I mean for the user, not for the person running the system. Make this one of your key metrics.

Regularly clear the decks of old systems and improve problematic ones. Appoint someone senior with accountability for continuous improvement of systems and processes, and add this to your incentive schemes and governance process.

Be careful not to add complexity in the pursuit of cost savings. Penny-wise can easily end up being pound-foolish. Examples of false economy include cutting corners on maintenance and eliminating production buffer stocks.

Remember that unit costs are your ultimate productivity metric, and the biggest impact on these is production. Attach a high value to stability and predictability, and recognise the importance of having flexibility within your operating system.

Target value, not costs. It is fine to target costs in the pursuit of value, but never let cost-cutting destroy value. This happens far too often in operations.

Simplification of your systems and processes should be one of your greatest business priorities. The simpler they are, the more likely they are to be understood and followed.

SOLUTION

The solution is to be really clear on your business strategy, your operating philosophy, and the boundaries of your overall operating system. Then be clear on everyone's role within this, not just within their own areas, but also in the interfaces with other areas. In particular, be clear how value is won and lost, in the parts, the interfaces, and most importantly in the way it all comes together to optimise the value of the operating system as a whole.

Then define the systems and processes you need to deliver this, and align them around delivery of this goal. Clear the decks of all other clutter and distractions. Make the systems as user-friendly as possible for the people using the systems.

As simple as that. This sounds obvious, but it is far from what is done in most businesses.

CHECKLIST

- How well is your 'whole operating system' defined? How well do your leaders and teams understand this whole system, their role within it, and how their area interacts with others around it?
- Do you periodically formally check how good this understanding is through your leadership team?
- How user-friendly are your business systems and processes? How would they rate on a bureaucracy scale?
- When you introduce new systems and processes, how much attention do you pay to simplicity and intuitiveness, from the user's perspective? And do you remove the old systems at the same time?
- When did you last do an overall review of your systems and processes (i.e. assess what you have; understand what works and what doesn't; then remove, improve, and redesign accordingly)?
- When you purchase new systems, do you pay the extra for the most user-friendly version, or do you try to save a buck by going for something that is a little more difficult for the user to interface with?

2.3 BEHAVIOUR IS MORE SILOED THAN COLLABORATIVE

OBSERVATIONS

Siloed behaviour is a constant negative force in the mining industry. It manifests as people and teams looking after their own interests above that of the overall organisation.

It means a team is not pulling together, just like a soccer team made up of a bunch of individuals who all want to be the star. It leads to selfish behaviour, wasted effort, and friction between players. There may be flashes of great performance, but in the long run it leads to suboptimal performance. A group of decent players working together tightly as a team will always beat a disjointed group of individual stars.

Siloed behaviour occurs everywhere in most organisations, and it's certainly prevalent in the mining industry. The following are some common examples.

Between business units and between mines: At a senior level, there is usually an awareness of the need for sharing learnings; however, in reality there is limited take-up of such ideas from other parts of the business. The barrier here is the fact that senior leaders have their own agendas and priorities, and there is a certain sense of 'not invented here' that stands in the way.

Between planning and execution: This is usually a significant gap in operations. The planners plan; the production people execute. Each blames the other for poor performance against plan: 'The plans aren't realistic' versus 'The production guys don't follow the plans like they should'. The real issue is typically that communication is lacking between them, they are pulling in different directions, and are focussed on different metrics of success.

Between production and maintenance: This is one of the clearest examples of siloed behaviour. The conflict here is that production teams have to deal with the uncertainty of unplanned breakdowns, and maintenance teams need the certainty of planned maintenance schedules. If planned maintenance is regularly put off due to breakdowns, a vicious circle is created, because you end up with more breakdowns and less planned maintenance. This conflict usually ends up with maintenance insisting on an absolutely fixed calendar-style maintenance schedule, which is seldom the smartest approach from an overall business-value perspective.

Between mining and processing: This is another common example of siloed behaviour. The nature of these two parts of the operation are very different. Mining is largely a noncontinuous process with multiple separate parts, where the value driver is volume and operating cost. Processing is usually a connected and continuous process, where the value driver is process efficiency and quality. A typical example of siloed behaviour is around fragmentation in, for example, a copper operation. The mining team is driven to blast their material as little as possible (to minimise costs) without affecting the flow through the crushers and stockpiles. The coarser feed constrains the throughput of the processing team, who are limited by plant capacity and would benefit from a much finer blast. Crushing and grinding is much more expensive than blasting, and the overall throughput impact can be as high as 10 percent, yet this critical performance driver is often not managed with the diligence it deserves.

Siloed behaviour is common, and typically leads to sub-optimal value of the overall system

There is more focus on the parts of an operating system ...

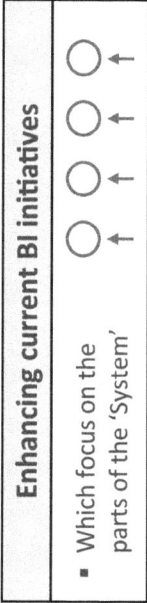

Enhancing current BI initiatives
▪ Which focus on the parts of the 'System'

BI = Business Improvement

- Businesses are typically structured managerially like this
- It is certainly easier to manage this way
- But it does lead to siloed behaviour, which can destroy value

Than on the interfaces and the whole

By targeting new untapped value
▪ Focus on the interfaces
▪ And the 'Whole'

- There is usually lack of transparency in what happens across the interfaces
- Businesses are often not well set up to deal with the dynamic changes which occur constantly in any operation (breakdowns, unavailability of resources, weather, etc)
- And very often, there is no clear responsibility, mandate, or even process for optimising the overall system value

The current spectrum across the industry

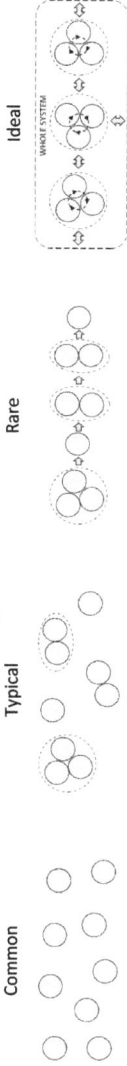

The circles indicate the various parts of the operating system

Fig 2.3 Behaviour is more siloed than collaborative

Even between adjacent shifts: Every production unit can point to the shift leader who produces more tons than the others. This is sometimes due to particular techniques (that may or may not be entirely desired), which are generally not openly shared with peers, because this point of differentiation is a point of pride and advantage when it comes to financial reward. Often this shift will have the reputation of pushing things so hard that they leave a mess for the next shift to deal with. This creates aggravation with the other shifts, especially when the behaviour is rewarded. The difference in production rate is sometimes quite significant, certainly in excess of 10 percent.

These examples highlight just a few of the big gaps and opportunities for improving performance, which are currently resulting in lost value. There are many more examples. None of these drive a culture of steady, continuous, standardised, predictable performance. They drive the exact opposite, yet we tolerate them every day as if they are acceptable.

There is clearly a need for a more collaborative and integrated approach, which is focussed on optimising the value of the overall operating system rather than the individual parts, and one that is less dependent on the style of the particular leader in charge at the time.

PRINCIPLES

Siloed behaviour is a natural phenomenon in organisations. It is nothing to be surprised about. It is much easier to work within your own boundaries than to have to worry about how your work fits in with other parties who have different views and priorities.

Anyone who has tried to work with other parts of a business to help you solve a problem will attest to the fact that it is easier said than done. Anyone who has tried to share successful ideas with colleagues in other parts of the business will recognise how surprisingly difficult it is to generate more than polite interest. Certainly, there isn't a natural inclination or 'pull' to share and replicate. Even joint problem-solving of common problems is hard.

There are many different drivers behind this—different viewpoints, different styles and approach, pride, arrogance, jealousy, ignorance, ambivalence—but perhaps most commonly, it is usually simply harder to work with others than to do it yourself. You need to consider others' views, which you may not agree with, and this will likely make it harder than you had originally envisioned.

The fact is that siloed behaviour is a common occurrence and is probably the natural default behaviour if left unaddressed.

So, collaborative, integrated behaviour is something that needs to be consciously managed, pursued, and driven. It is not something that will just happen by itself. It is one of the key behaviours that needs to be developed within an organisation if it aspires to achieve operational excellence and maximise the value delivered by its operations.

SOLUTION

What is required is much more than just training around collaborative behaviour.

Integrated, collaborative behaviour needs to be embedded into the DNA of the business, into all of the management processes, into the way information is communicated, into establishing an effective working environment for collaboration, into integrated systems, and into an organisational architecture that encourages, enables, and drives integrated, collaborative behaviour.

Not only must siloed behaviour be discouraged, it must be almost impossible to occur due to the nature of the systems and processes embedded into the business.

CHECKLIST

- Is your planning fully integrated, or is it largely done in silos and then patched together through a collaborative process?
- Do your production and maintenance teams develop separate plans/schedules, or a single plan/schedule that incorporates all activities?
- What systems, processes, and accountabilities do you have for targeting value lost in the interfaces between teams?
- What systems, processes, and accountabilities do you have for optimising the value of the whole operating system?
- Do your incentive schemes reward individual performance or team performance?
- How do you define your 'teams'? Production, maintenance, combined? Shifts, sections, departments?
- How is your workforce trained in the behaviours expected of them in managing the operations in a collaborative, integrated way?

2.4 SENIOR LEADERS NOW FOCUS TO MUCH ON THE SHORT TERM

OBSERVATIONS

As you move up through the ranks, your role should become less about the short term and more about the long term. At the executive level, the role should be a reasonable balance between shaping the vision and aspiration (20 percent), crafting the strategy (20 percent), building capability in the team (30 percent), and managing performance (30 percent).

Unfortunately, this is seldom the case today. For too many executives this has now evolved to become:

- Shaping the vision and aspiration (0 percent)
- Crafting the strategy (5 percent)
- Building capability in your team (5 percent)
- Managing performance (90 percent)

For various reasons, managing performance has become the dominant focus of many executives and senior leaders. This has been largely driven by technology changes, which have increased the intensity of, and accessibility to, data and information. Corporate centres are becoming more involved in the decisions made at their various operations, and individual mines certainly have much less autonomy than they used to have twenty to thirty years ago. This has changed the nature of the pressures on local business units. Today, short-term results rule. Long-term results are for the next person in your role to worry about.

It's not that long-term sustainable results are not sought after; it's just that they are much harder to deliver, because they involve convincing and aligning with many others. It is hard enough just to get your own immediate challenges fixed.

This has the effect of pushing everyone's roles down in the organisation. When senior leaders start micromanaging, it sends the message that they are running the show, and their teams start to sit back and wait for the next instruction. It is fine for leaders to question the detail, but this needs to be done in a way that drives accountability and ownership of the problem and solution, down through the line.

As a senior leader, if you ask, 'Why was that piece of equipment down so long?', 'Whose fault was it?', and 'What are you doing about it?' you quickly get sucked into a finger-pointing exercise and become part of solving the problem. If you ask, 'Was that equipment downtime due to poor planning, or was it due to poor execution of the plan?' and 'Are you going to take steps to prevent it from happening again?' and 'When will those steps be completed?' you stay out of the problem-solving itself, and sharpen the accountability of the person you are talking to. The moment you get sucked into the problem-solving itself, you take ownership away from where it should be held, and that pushes their accountability down a level.

In today's operations, the emphasis is too often on dealing with specific, immediate problems and short-term solutions. This inevitably leads to short-term behaviour and short-term results. It detracts from the true underlying problem, which is constant fire-fighting and the lack of predictable performance and systemic, sustainable improvement.

The role of an executive has changed over the past 20-30 years

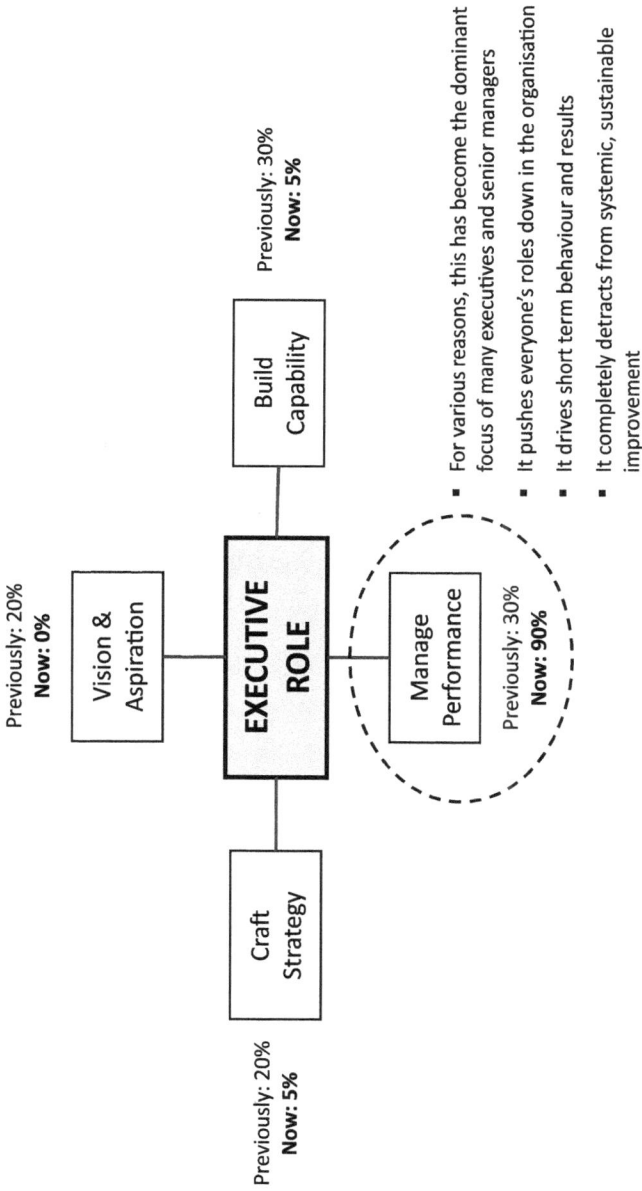

Vision & Aspiration
Previously: 20%
Now: 0%

Build Capability
Previously: 30%
Now: 5%

Craft Strategy
Previously: 20%
Now: 5%

EXECUTIVE ROLE

Manage Performance
Previously: 30%
Now: 90%

- For various reasons, this has become the dominant focus of many executives and senior managers
- It pushes everyone's roles down in the organisation
- It drives short term behaviour and results
- It completely detracts from systemic, sustainable improvement

Fig 2.4 Senior leaders now focus too much on the short term

Systemic solutions have a lot more to do with the way the operation is run, how the different teams work together, and how well everyone understands the overall game plan and their role within it. Systemic solutions also involve the clarity and simplicity of the systems and processes being used, the training and capability of the teams, and how well both good and bad performance are recognised and followed up.

PRINCIPLES

Most operational teams are good at what they focus on. If the focus of discussions is mainly on short-term issues, you can expect behaviour that drives good fire-fighting and impressive short-term results. If you want long-term, predictable performance and continuous, sustainable improvement, you need to focus your discussions on the things that drive this behaviour. They are not the same.

The industry is currently quite good at fire-fighting. It is what distinguishes the good managers from the bad, to a large extent. Businesses are great at identifying and actioning long lists of things that are wrong. Consulting companies have developed tight systems to manage literally thousands of 'high-value', 'quick-win' initiatives. Operational teams get consumed by the pressure heaped on them to address this multitude of projects.

The problem is, this long list of quick-win projects collectively somehow never quite produces the overall results expected. The reason for this is that they tend to address the symptoms, not the root cause. The various problems and solutions are mostly intertwined to some degree, and so, solving one problem often creates another in another area. If the focus is on 'ticking the boxes' on your long list of projects, that is what you will get. It won't address the underlying, systemic issues, and you most likely won't get the overall lift in performance you expect.

It is important for senior leaders to increase their focus on the long term, as well as on solutions that address the root cause and are sustainable. Not only does this make good business sense, but it also sends a positive message to their teams that their leaders care about improving the working environment and making it easier for everyone to succeed.

SOLUTION

Business leaders need to recognise that this is a big gap in the way they are currently leading and managing their operations.

This warrants some initial exposure for the senior leadership to the nature of the gap that exists, and the need for a different emphasis. An engaging, codesigned approach is the best way of achieving this with leadership teams, to ensure deep understanding and ownership of any changes involved.

The answer itself lies in identifying the systemic underlying issues behind the problems and focusing on *them*. This doesn't mean you should throw away your long list of improvement opportunities; it just means you tackle them in a different way—in an integrated way, with an underlying holistically designed approach that addresses the usually multidimensional nature of the cause.

This requires a multidimensional lens (the lens of integration), which centres around five key elements: management process, transparency of information, collaborative environment, integrated systems, and organisational architecture.

The organisational architecture piece of this is a crucial part of the design. This involves redefining the roles, processes, and expectations around planning and execution. There needs to be greater distinction between these two functions, with commensurate increased accountability to deliver realistic plans, and to follow those plans in a disciplined way.

CHECKLIST

- How would you rate your senior leaders' balance between: shaping the vision and aspiration (? percent), crafting the strategy (? percent), building capability in your team (? percent), and managing performance (? percent). Better still, how would the subordinates each rate their leaders in this regard?
- How many of your improvement initiatives are really addressing the long-term cause of your problems, rather than just the short-term symptoms?
- How efficiently and effectively are the improvement initiatives being completed, and how does this correlate with overall performance improvement?
- What is your current emphasis on, and approach to, identifying and addressing underlying systemic issues?
- Do these consider the usually multidimensional nature of the underlying cause?
- Do you have a structured process for identifying and addressing such systemic issues?

2.5 CURRENT BUSINESS IMPROVEMENT MODELS ARE NOT WORKING

OBSERVATIONS

The business improvement (BI) goal of many mining companies today involves trying to get back to productivity levels of fifteen to twenty years ago. This is a poor indictment of the state of business improvement in the industry generally.

There are many problems with the current BI model used in the industry. The broader of these are covered by the underlying message of this book, which is that the importance and role of integration has been underplayed. However, more specific issues are explored below.

'Big stick', nonsystemic approach: When a new manager starts in the role, they typically inherit some problems that they have been brought in to solve. The usual approach is to look thoroughly at what is happening across their whole area of control. Then they use judgement to decide what are the top, say, three priorities. They clear the decks of all other distractions, move a few people around, and drive those three priorities with the aforementioned 'big stick'. And it works. Things improve, and this is the way most successful leaders prove themselves and are moved on to another role. The problem with this is that the replacement manager then arrives with a new challenge, and they follow the same approach. But the original three problems have been fixed, so the focus is on another three issues. Same approach: 'big stick' focus on the new three, and all other distractions removed. Again, these three areas improve. However, the original three issues that were fixed through the 'big stick' approach, not through anything systemic and sustainable, now pop up again. And so the cycle is repeated over and over again. Improve, decline, improve, decline.

Short-term improvement metrics: Improvement targets are usually quite specific in nature, and are set against a baseline—normally last year's results. So, assuming you have a really bad year, and then improve the following year, you can decline the year after and improve the year after that, and actually hit your improvement goals 50 percent of the time, without going anywhere in terms of real sustainable improvement.

Cost versus value: In a mining environment, production tons are hard to improve without changing the equipment capacity of the system. It is even harder to pin any improvement on the specific actions you have taken, given the multitude of variables involved. So, there is a tendency to focus a lot on costs rather than less-tangible metrics like productivity and efficiency, which would be better measures of success. Directives from senior management are often quite specific about cost-saving targets. There is an obvious logic (and necessity) behind this; however, it can drive a very dangerous behaviour. Given enough pressure, a manager will just find a way to comply somehow, to keep their job and take the pressure off themselves. This leads to things like high-grading the ore body, cutting production buffer stock flexibility, and delaying things like sustaining capital and maintenance. Each one of these examples happens all the time across the industry, and it very often leads to major cost/production problems later on. Short-term gain, long-term pain.

Big programmes and long lists of projects: Big improvement drives, particularly when they take the form of a big-branded

There is currently not enough focus on systemic, sustainable improvement

- There is too much focus on lists of 'improvement initiatives', which address the **symptoms** not the cause

- These usually have a short-term impact, because they are **non-systemic** in nature and don't address the root cause

- The resulting improve-decline-improve-decline cycle leads to constant reinventing of the wheel

- This is a **poor platform** for operational excellence and a progressive technology strategy

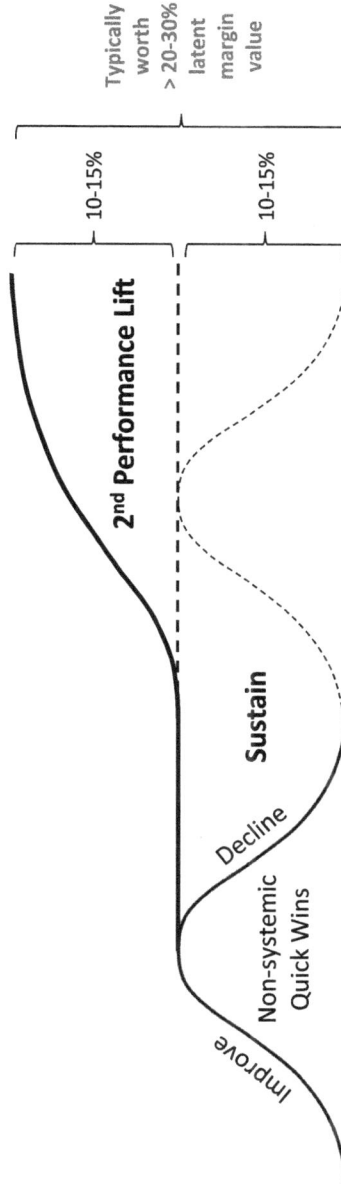

Typically worth > 20-30% latent margin value

10-15%

10-15%

2nd Performance Lift

Sustain

Decline

Non-systemic Quick Wins

Improve

Fig 2.5 Current Business Improvement models are not working

programme of some sort, typically lead to huge lists of improvement projects. It all starts well, with the operational teams asked to list all their problems and the potential impact. Then value is attached to each of these, and the central organiser starts to put the heat on those same operational people to fix the problems (the same problems they have never been able to fix in the past!). Then, when the improvements and quantifiable value are slow in coming, the heat is increased, and the 'big stick' comes out again. Eventually the operations staff are forced to come up with something, and it is often non-value adding, and/or distracts them from other core management priorities like time out in the field and sometimes safety. It is common for such programmes to generate hundreds of specific initiatives, and in big companies literally thousands, with mind-blowingly complex spreadsheets used to track the huge amounts of information involved. A consultant's dream! The cost-saving number attached to these 'successful' programmes is usually very substantial, but there is often a severe price paid later on.

Such programmes and such efforts have been tried for decades, yet they have not materially moved the dial, or resulted in a significant impact on long-term productivity or improvement culture within the industry.

The current BI model is not working effectively, and it is time to look at things differently.

PRINCIPLES

The common problem with all of these examples is that they all drive short-term solutions and short-term results, which at best do nothing for continuous, sustainable improvement, and at worst can be hugely value destroying.

The lack of sustainability in the current improvement efforts means that businesses are destined to continuously reinvent the wheel and operate in an unstable mode. And very importantly, this is a poor platform for operational excellence and a progressive technology strategy, which means you will always struggle with the next performance lift in the future.

Addressing the fundamental cause of problems makes a lot more sense than constantly addressing the symptoms. It is because it is difficult to identify what the fundamental cause is, that this is generally not well done. Most BI teams will say they try to get to the root cause, but in reality, they seldom do. Their 'root cause' too often just includes another dimension or two, which is a good start, but by no means the full picture. The true root causes usually involve addressing some deeply embedded constraints, which lie well beyond the control or influence of the respective manager. So they stick to their own domain, and thereby don't end up dealing with the fundamental systemic issues.

In order to truly define what needs to be done to achieve fully integrated solutions and sustainable results, you need a structured approach and a desire to fix things properly rather than quickly and superficially. A good mindset for BI is to have as your mantra, 'An improvement isn't an improvement unless it is systemic and sustainable'.

Other aspects to reflect on:

- Focus on making the workplace easier for your people to succeed in.
- Simplify systems and processes.
- Culture is not achieved through a training course. It is an outcome when the behaviours you want are embedded into your systems and processes, and are fully aligned and consistently applied in practice.
- Improvement goals and incentives should not all be centred around project delivery; they need results.
- Metrics and incentives for BI teams should centre around continuous improvement trends, simplification of systems and processes, and achieved standardisation—not completing studies, achieving project milestones, and writing standards.

SOLUTION

By all means start with a list of all the issues and opportunities in each area.

But then you need to have a higher-level collective workshopping of what the burning platform issues are in the business. Collect and compile evidence of each of these problems, and identify the underlying reasons and changes needed for each one. But then, instead of jumping to actions, you need to look through all of these collectively from a different angle. Look through the lens of integration, which will help you identify the systemic, underlying issues.

Then you are in a position to identify the systemic solutions and the best way of addressing and resourcing these. They will be relatively few in number.

Leadership and clarity of purpose are important in all of this. Ensure everyone is playing to a single game plan and they each understand how their role contributes to that. Ensure everyone is committed to getting better and better at what they do and making it easier for others where they can.

The key to unlocking this is the lens of integration and an intent to properly solve the core problems in the business rather than just achieve quick-win solutions. A short-term view may take the pressure off and look good for a while, but it will just pass on the same problems for your successor to solve again in the future.

CHECKLIST

- How many improvement projects do you have in total across your business?
- Does that sound like a reasonable number to you, and does it give you a feeling of confidence that they are all well-aligned and integrated, and are collaboratively addressing the underlying systemic causes?
- How many of these are focussed on systemic solutions and simplification, and making it easier for your teams to do their jobs efficiently and effectively?
- Are your BI teams incentivised to deliver continuous, sustainable, long-term improvement, or are they primarily rewarded for completing projects and milestones?

THE OPERATIONAL BASICS NEED TO BE TAKEN TO A WHOLE NEW LEVEL

//

3.1 Focus On Optimising the Whole System, Not Just the Parts

3.2 Systemically Improving the Quality of Planning, Execution, and Improvement

3.3 Fully Integrated Planning

3.4 Disciplined Execution

3.5 Continuous, Sustainable Improvement

3.6 Hard-Wiring Safety and Risk into Your Routine Management Processes

3.7 Leveraging the Value of Data

3.8 Future-Proofing the Changes

3.1 FOCUS ON OPTIMISING THE WHOLE SYSTEM, NOT JUST THE PARTS

OBSERVATIONS

As already discussed, siloed behaviour destroys value. Focussing just on improving the parts of an organisation does not deliver the best result for the business as a whole.

The goal in any operation should be to align everyone and everything around delivery of the business strategy, and to avoid any distractions and dilution from that effort. In essence, this is what integration is.

If you consider a business to be made up of several parts (for example, the different departments making up the value chain), this means optimising the parts, optimising the whole, and doing this in a way that facilitates integration between the two. Let me further explain.

Optimising the parts: This means managing each part of the operation professionally, to ensure reliable and predictable operations. Functional excellence is important here, and this is an area of concern for many operations, as the depth of knowledge and expertise can become diluted over time. Also, it is important that the focus is not simply centred on maximum production all of the time, as this invariably leads to interruptions like breakdowns and blockages. The goal should be informed, predictable, sustainable, and aligned with the other parts of the operation.

Optimising the whole: This means optimising the value of the operating system as a whole. This is typically a gap in operations,

Integration is about pulling everyone and everything together in the same direction (ie around delivery of the business strategy)

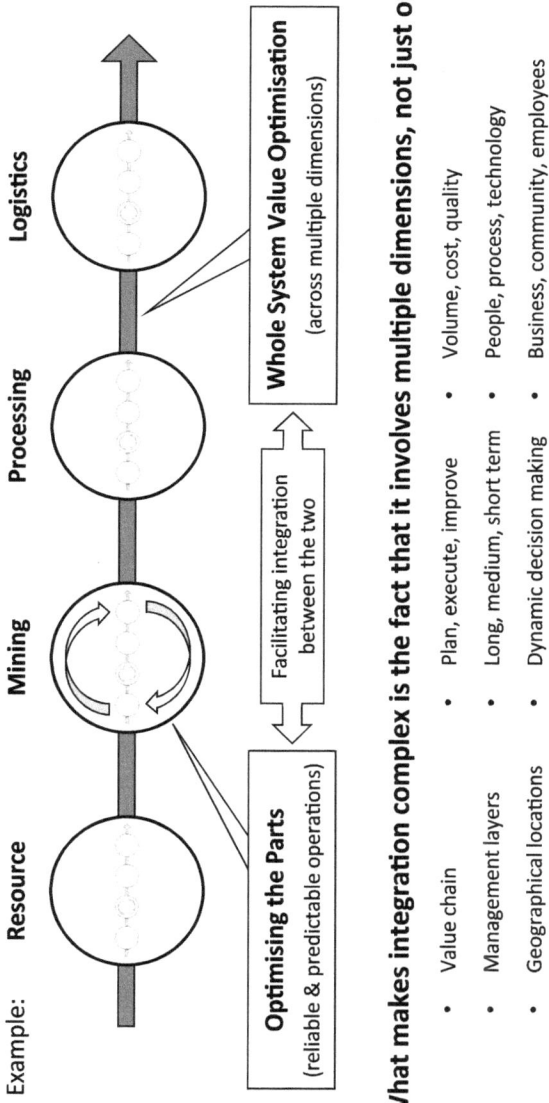

Example:

| Resource | Mining | Processing | Logistics |

Optimising the Parts
(reliable & predictable operations)

Facilitating integration between the two

Whole System Value Optimisation
(across multiple dimensions)

What makes integration complex is the fact that it involves multiple dimensions, not just one

- Value chain
- Management layers
- Geographical locations

- Plan, execute, improve
- Long, medium, short term
- Dynamic decision making

- Volume, cost, quality
- People, process, technology
- Business, community, employees

A consistent, structured and holistic approach to doing this is therefore essential

Fig 3.1 Focus on optimising the whole-system, not just the parts

as there is no one accountable for optimising the whole operating system. There may be a technical function that oversees the whole system, but this doesn't normally extend to a mandate and full authority to make whole-system decisions. The senior role (e.g. GM) obviously has responsibility for the whole system, but as a single individual, especially in an inexperienced management team, they are not able to stay connected to all of the detail that such a role requires. Therefore, the strongest voice in the value chain typically wins the argument for their area, which is not always the best result for the business as a whole.

Facilitating integration between the two: This means optimising the parts and the whole in a way that makes it easier for each to talk to each other. For example, in operations, the mining department is typically set up very differently from the processing department—different structures, different systems, different management processes, different metrics, different agendas, and different culture. So, it is not surprising that these two departments often find it difficult to pull in the same direction.

What makes this so difficult: The difficulty comes from the fact that the value chain is just one of many dimensions that need to be integrated. And, of course, even the value chain has multiple levels (e.g. drill, blast, load, haul within a mining department; resource, mining, processing, logistics within a mine; mines one, two, and three within a business unit (BU); and BUs one, two, and three within a corporation. However, integration is also required across the different management levels; across different geographical locations; across planning, execution, and improvement functions; across the long-, medium-, and short-term; and between operations and support functions. Also between volume, cost, and quality value drivers; between people, process, and technology, etc. When you consider all of these dimensions, you quickly realise that every individual in an operation has a different combination of these. Therefore, this is what pulls everyone and everything in slightly different directions.

This is why you need to have a structured approach to simplify this process. This structured approach is an integrated operating model. What is needed is a clear understanding of what integration actually means, and a common 'lens' of integration, which allows you to align all of these aspects systematically.

PRINCIPLES

The structured approach starts with getting very clear on the overall business strategy, the operating philosophy, and the definition of the whole operating system.

Then you need to focus your efforts on all three components required to achieve integration: optimising the parts, optimising the whole, and facilitating integration between the two. And you need to consider all of the different dimensions highlighted earlier.

Sound complicated? Questioning why you would go through all of this? Perhaps ask yourself this: 'Why would I not want to get really clear on the game plan for the business, and align and cascade what everyone's role is within that? Then lock this into systems and processes that make it easier for them to work together to achieve this?'

The sad fact is that most businesses do nothing of the sort, to this level of rigour. As a result, everyone has a slightly different view of the game plan and their role, and they work within silos, with access to slightly different information, within slightly misaligned management processes, and to slightly different agendas. And we wonder why value is lost in all the multitude of interfaces that make up a complex business.

Yes, it does require some hard work up front to do this properly. But doesn't it make sense to get everyone clear and aligned up front? When you think about it, how can we possibly justify doing anything else? Once you understand this in these terms, it just seems irresponsible to even think about being vague about all of these things, and allow significant degrees of freedom at each and every stage and interface, which just contributes to an overall outcome of looseness, misalignment, fragmentation, and wasted value for the business.

This is precisely where most operations sit at the moment. The lost value attached to this is enormous, and the frustration caused by it is considerable.

SOLUTION

The solution is to take a step back and really think about the way your business is set up—from the perspective of overall business value, as well as from the perspective of everyone working within it.

Does it really need to be as complicated as we make it? How much wasted energy and value is lost because the left hand doesn't know what the right hand is doing? How much time and value is wasted trying to find information, arriving at one version of the truth, debating pros and cons between the different parts of a business affected by a problem, and then getting someone to make a decision?

How much better would it be if production and maintenance worked together around a common agenda, instead of disrespecting what each other needs in order to deliver a quality service to the other? How much easier would it be for operations if support functions were totally dedicated to making it easier for the people in the front line to do their jobs?

An integrated approach is actually a simpler approach. It causes less friction, less frustration, better alignment, and improved value.

CHECKLIST

- How well-integrated is your business across all of these dimensions?
- Have you tried to quantify or benchmark integration?
- Instinctively, what sort of value do you feel might be attached to areas that are not currently well-aligned?
- Are you currently incentivising the behaviour you want/need to drive (i.e. overall system performance)?
- Do you believe that your systems and processes could be simplified for the people using them?
- What is holding you back from addressing these issues?
- Who would be the right person in your organisation to drive such a change?
- Given the value attached to this, why would this not be a full-time role?

3.2 SYSTEMICALLY IMPROVING THE QUALITY OF PLANNING, EXECUTION, AND IMPROVEMENT

OBSERVATIONS

What we should be aiming for in operations is a well-focussed and integrated approach, with a healthy balance between outstanding planning, execution, and improvement.

Planning: Outstanding planning means it should be optimised, integrated, dynamically managed, and realistic. The focus of this should be all about ensuring it is well-informed and optimises whole-system value. Typically, operations are far from this. The norm is for siloed and poorly integrated plans, with disconnects between the long and short term, and with poor management of dynamic changes.

Execution: Outstanding execution means it should be about disciplined execution of the plan, with dynamic changes addressed in a whole-system-value context. The focus should be on steady, predictable performance rather than boom and bust. Typically, operations are far from this. The norm is for a firefighting approach, where short-term priorities win, there is poor conformance to plan, and dynamic decisions are handled in a siloed way.

Improvement: Outstanding improvement means a focus on continuous, systemic, sustainable improvement. The focus of this should be about simplifying and standardising the workplace and embedding this into a high-performance culture. Once again, typically, operations are far from this. The norm is that systemic improvement is an afterthought, and the focus is on short-term, point-solution improvement initiatives, which are focussed on

Taking planning, execution, and improvement to a whole new level

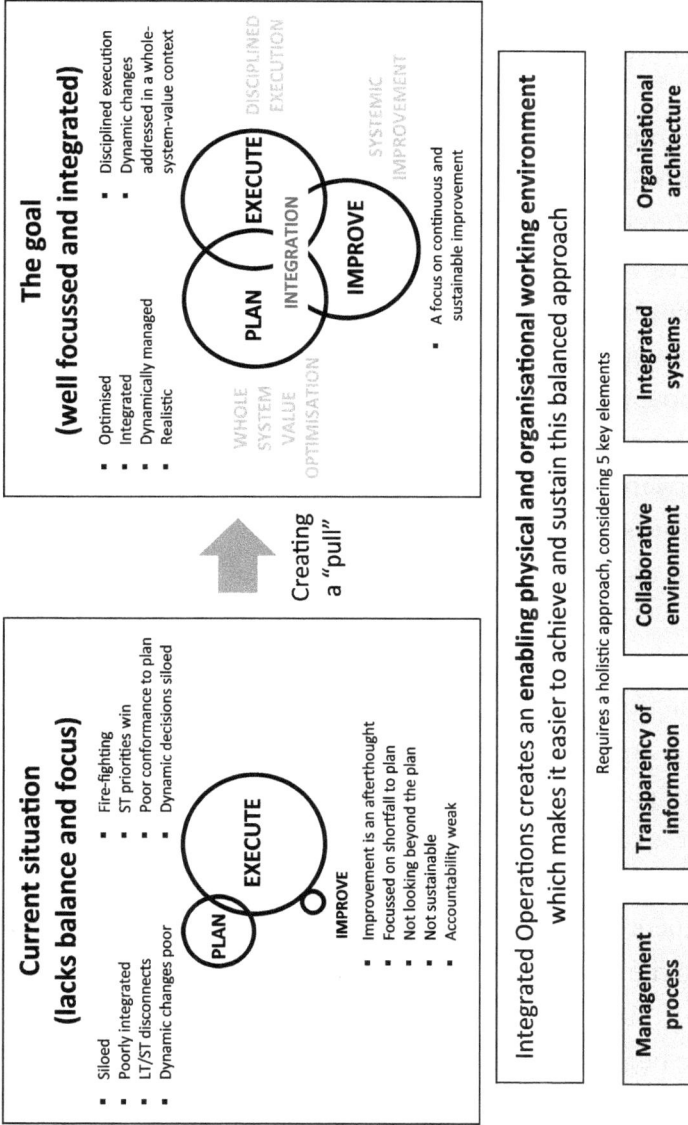

Current situation
(lacks balance and focus)

PLAN
EXECUTE

- Siloed
- Poorly integrated
- LT/ST disconnects
- Dynamic changes poor
- Fire-fighting
- ST priorities win
- Poor conformance to plan
- Dynamic decisions siloed

IMPROVE

- Improvement is an afterthought
- Focussed on shortfall to plan
- Not looking beyond the plan
- Not sustainable
- Accountability weak

Creating a "pull"

The goal
(well focussed and integrated)

WHOLE SYSTEM VALUE OPTIMISATION

- Optimised
- Integrated
- Dynamically managed
- Realistic

DISCIPLINED EXECUTION

- Disciplined execution
- Dynamic changes addressed in a whole-system-value context

PLAN **EXECUTE**

INTEGRATION

IMPROVE

SYSTEMIC IMPROVEMENT

- A focus on continuous and sustainable improvement

Integrated Operations creates an **enabling physical and organisational working environment** which makes it easier to achieve and sustain this balanced approach

Requires a holistic approach, considering 5 key elements

| Management process | Transparency of information | Collaborative environment | Integrated systems | Organisational architecture |

Fig 3.2 Systemically improving the quality of planning, execution, and improvement

the shortfall to plan rather than looking beyond the plan. The efforts are usually characterised by a long list of projects, with weak accountability for the BI teams, and any improvements are seldom sustained.

Systemic working environment: A systemic working environment means creating an enabling physical and organisational working environment that makes it easier for your teams to achieve and sustain this balanced approach. The key to this is to focus on creating the enabling environment, which creates a 'pull' towards continually driving that behaviour. It should not be about just driving production flat out regardless of how, and the planners struggling to keep up with all the changes, and the improvement team chasing after a long list of discrete project initiatives to try to make up the shortfall. Unfortunately, this is the all-too-common scenario.

Priorities: To illustrate this point, generally speaking, most of a production meeting is usually taken up with immediate 'execution' issues and how to resolve them. There is some discussion about 'planning'—otherwise you wouldn't get anything done—but it is often reactive, rushed, and not fully thought through. Then at the end of the meeting, when time has run out, comes a comment by the chairperson to the 'improvement' representative along the lines of, 'Sorry, Jo, I know you have some really interesting things you wanted to discuss. We'll give you five minutes at the beginning of the next meeting'.

PRINCIPLES

The key is for the planning to be informed and optimised for whole-system value, the execution to be disciplined and predictable, and the improvement to be systemic and sustainable. These three components must also be well-integrated.

Planning principles: Stretch planning is a commonly used approach, but it has serious drawbacks. First, it confuses the picture. When you ask an operations leader how they are doing compared to plan, you often get an answer along the lines of, 'Which plan? The weekly plan? The monthly plan? The annual plan? The original budget? The revised budget? The forecast? The revised forecast? The "stretch plan"? The target?' Second, by definition, a stretch plan is slightly unrealistic. And an unrealistic plan can quickly become viewed as such, and be cast off by operators as someone else's impractical wish-list number, which just dilutes accountability for delivery. A realistic plan increases accountability and does not dampen the inherent human desire to exceed expectations. It also removes pressure for operators to be trying to break records all the time, which usually leads to inconsistent and unpredictable performance. Less haste, more speed. Consistent, steady, predictable performance wins in the long run.

Execution principles: If the plan is optimised, integrated, dynamically managed, and realistic, then the role of the production team should simply be to execute the plan, with as high a degree of compliance as possible. Any dynamic changes (e.g. breakdowns or weather events) should be handled from the perspective of what is best for the overall operation, not what is convenient for the team where the problem or change occurred. If multiple teams collaborate to produce a single, integrated plan and schedule, then compliance to that plan and schedule should be the driving metric. If this is backed up by a well-structured and mandated process for continually managing dynamic changes from a whole-system-value perspective, then you have a solid basis for achieving disciplined execution.

Improvement principles: The 'execution' team should be managing a certain degree of variability themselves—certainly with respect to day-to-day variations and issues that arise, and making up shortfalls from the plan (which, by the way, should have been set up to be realistically achievable). Making up production shortfalls is not the job of the 'improvement' team. Their role should be to drive continuous improvement, systemically and sustainably, through things like smarter systems and processes, sharing leading practice, standardisation, and simplification. They should be measured by long-term improvement trends in key metrics (including compliance to plan/schedule, variability, predictability, etc).

Organisational principles: A big part of the reason for the problem of unbalanced priorities is the lack of separation between accountability for planning, execution, and improvement. If accountability for all three is given to one person, the short-term urgent problem will always take priority over the long-term important. This isn't a criticism; it is simply a reality, because long-term improvement is a moot point if you don't survive the short term. This is the reason why the 'Current Situation' circles look like they do in Figure 3.2. Therefore, it is important to improve the separation between 'execution' and 'planning and improvement'. This needs to be done with care, and there are some important tricks to getting this right. One of these is that the 'planning and improvement' function must have equal accountability for production results, so this doesn't just heap more pressure on the 'execution' roles.

Integration principles: Once the planning, execution, and improvement teams are truly working in sync to a common game plan and aligned metrics, it makes everything much simpler for everyone—a well-oiled machine, working together constantly as a team. It should be a systematic process, which does not rely on a constant 'big stick' to drive it. Common sense, really.

SOLUTION

As already mentioned, to achieve this integrated and aligned approach, you need to consider five key elements, holistically. These are: management process, transparency of information, collaborative environment, integrated systems, and organisational architecture.

You then need to align the physical and organisational working environment holistically around these elements. Each individual element doesn't have to be complex (in fact, it is better if it is kept simple), but all of the elements do need to be aligned with each other.

For example, if you change the KPIs (key performance indicators) for the business, you can't just communicate this to the team. You need to embed it into the meeting processes, reporting systems, and the way you communicate the plan and performance against the plan. You need to make sure this is aligned across the organisation and cascaded down through all levels. You need to introduce real-time data tracking, analyse this data, and feed this back into a decision-making process with clear accountabilities for making decisions around this. You need to incorporate all of this into your training systems and processes, including induction and refresher training. You need to adjust your incentive programmes where necessary. And you need to engage your leadership team in ensuring this message is understood and consistently practiced effectively, right down through to the front line. This should be periodically surveyed and audited to ensure it is sustained.

Then you have a sustainable system. Does this sound impractical? Why would you consider it to be impractical? What is the alternative? Staying with the current way of doing things? Is that any easier? I don't think so.

CHECKLIST

- How many plans, budgets, forecasts, and targets do you have running concurrently?
- Does this confuse frontline operations people? How could you simplify this to a single, live, dynamically adjusted view of what is expected of them during each shift, day, week, and month?
- How well set up are you for managing dynamic changes across the organisation (whether everyday internal events or periodic external shifts) in a way that optimises the value of the business as a whole rather than the individual parts?
- How well-informed are your plans? How predictable is your execution? How sustainable is your improvement?

3.3 FULLY INTEGRATED PLANNING

OBSERVATIONS

We have already discussed how the key for good planning is that it is well-informed, fully integrated, and optimised for whole-system value.

Integration plays a very important role in this, hence the emphasis on integrated planning. This implies that you need to consider all aspects that could affect the plan. You need to understand how all of this information interrelates, and you need to apply judgement in coming up with a single view that optimises for value (noting that value can take many forms and should consider both downside risk and upside opportunity).

Therefore, in order to properly integrate your planning, you need to consider the whole plan-execute-improve cycle. This is more than just a three-step process, and in a mining context, should look something like Figure 3.3 in practice. Other industries should look similar with just different examples in the detail.

The following is a further explanation of each of these stages and explores the challenges involved, using mining as an example.

- **Consider all options and shape the strategy:** This is important. Don't limit yourself to certainties at this stage. Consider scenarios and external factors. Consider the strategies that will define your business expectations, priorities, and assumptions. This should include opportunities, risks, options, and constraints. This is usually not done in a particularly structured way, so the assessment of opportunities, risks, and options is not always rigorously explored.

- **Develop a single, fully integrated life-of-mine (LOM) plan:** Based on the broad strategic assumptions, define the optimum way of mining the ore body, i.e. the LOM plan. This is the aspect of planning that is done best by mining businesses, because the block model requires it. It is the one place where information is extremely well integrated and value trade-offs are explored in a structured and objective way. An LOM plan is by definition a plan that is optimised for whole-system value.

- **Develop the specific annual and quarterly detail behind the LOM plan:** This is where the planning process starts to break down. In developing the annual and quarterly detail, the granularity requires more specific detail around the assumptions, which is not always clear at that stage. Therefore, as the years go by, it is inevitable that disconnects start to occur between the optimised LOM plan and the more granular detail of the annual and quarterly plans. In the absence of a strong process that is run by a team with all the information at hand, this typically becomes a rather clumsy process. It gets caught up in the budgeting cycles, and decisions often get made in haste or without full context, in a loosely coordinated rather than fully integrated and objective way. Connection back to the assumptions and commitments made in the LOM plan is not always done rigorously, and reconciliation processes are often weak and not as transparent as they should be.

- **Schedule the plan:** This is where the quarterly plans are detailed into monthly, weekly, daily, and shift schedules. Whereas the annual and quarterly plans are usually developed

Whole-system value-optimisation through integrated Planning and Improvement

- Integrated planning is not just about developing an integrated plan
- It is about considering all phases of planning, execution and improvement, across all time horizons
- It is a complex process, which requires ongoing iteration to ensure value is optimised in a continually changing environment
- An effective Integrated Planning function brings context, discipline, transparency and accountability to this process

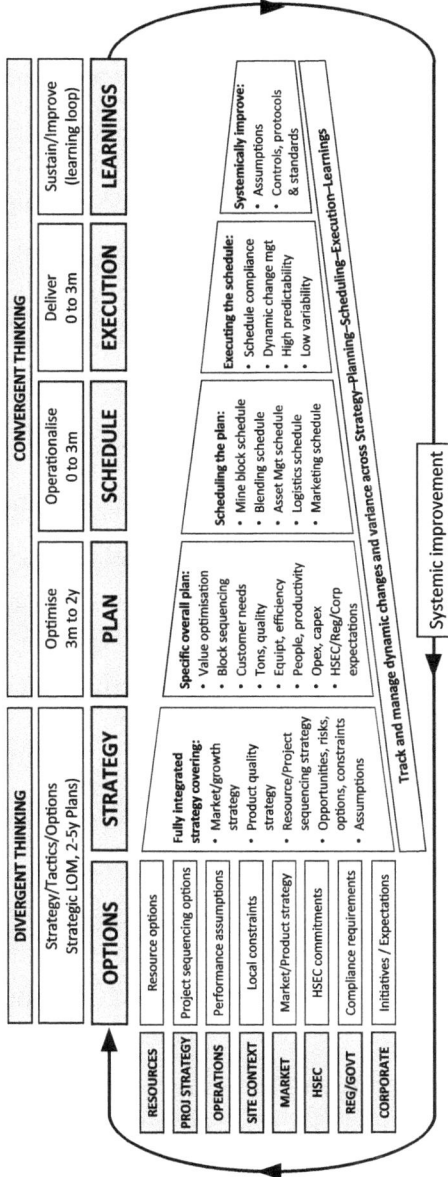

DIVERGENT THINKING		CONVERGENT THINKING		
Strategy/Tactics/Options Strategic LOM, 2-5y Plans	Optimise 3m to 2y	Operationalise 0 to 3m	Deliver 0 to 3m	Sustain/Improve (learning loop)
OPTIONS · STRATEGY	PLAN	SCHEDULE	EXECUTION	LEARNINGS

OPTIONS

RESOURCES	Resource options
PROJ STRATEGY	Project sequencing options
OPERATIONS	Performance assumptions
SITE CONTEXT	Local constraints
MARKET	Market/Product strategy
HSEC	HSEC commitments
REG/GOVT	Compliance requirements
CORPORATE	Initiatives / Expectations

STRATEGY

Fully integrated strategy covering:
- Market/growth strategy
- Product quality strategy
- Resource/Project sequencing strategy
- Opportunities, risks, options, constraints
- Assumptions

PLAN

Specific overall plan:
- Value optimisation
- Block sequencing
- Customer needs
- Tons, quality
- Equipt, efficiency
- People, productivity
- Opex, capex
- HSEC/Reg/Corp expectations

SCHEDULE

Scheduling the plan:
- Mine block schedule
- Blending schedule
- Asset Mgt schedule
- Logistics schedule
- Marketing schedule

EXECUTION

Executing the schedule:
- Schedule compliance
- Dynamic change mgt
- High predictability
- Low variability

LEARNINGS

Systemically improve:
- Assumptions
- Controls, protocols & standards

Track and manage dynamic changes and variance across Strategy–Planning–Scheduling–Execution–Learnings

Systemic improvement

Good integrated planning requires strong management processes, transparency of information, a collaborative environment, integrated systems, and the right organisational architecture, to be successful

Fig 3.3 Fully integrated planning

by the central planning teams, the shorter-term schedules are usually developed by the operations teams. For similar reasons to the planning process, this scheduling process typically introduces further disconnects between the original intent of the plan and what gets scheduled. Again, the reconciliation and governance processes are not as strong as they should be—exacerbated by the fact that, this time, different teams are usually involved, and actual equipment breakdowns, production shortfalls, and other issues start to play into the mix, which constrains the options open to the scheduling teams.

- **Execute the schedule:** This is the point where the process really breaks down. The schedules are usually either based on average information or out-of-date information. Because the schedules are not completely realistic, the frontline operators often use them as just a guideline to get the gist of what is expected and then apply various degrees of freedom to the way they actually do the implementation. This leads to deviations from yesterday's schedule, and this in turn necessitates rescheduling the next day, which ties the schedulers up with rework instead of refinement of the plan. This creates a vicious circle that leads to unrealistic schedules and poor execution-compliance. The fact that production, maintenance, and services teams schedule their work separately also results in further inefficiencies at the interfaces between the schedules. That's hardly a solid foundation for operational excellence.

- **Track and continuously manage dynamic changes and variances at every stage:** What compounds the problem even further is that dynamic changes (e.g. breakdowns and other constraints) are constantly occurring in every area (partly caused by the unrealistic scheduling), and there is usually no real structured system or process for dealing with these changes. This is a major weakness in most operations. Some operations have introduced an Integrated Operations Centre (IOC), which is a structured way of managing these dynamic changes. For some operations, this is a complex facility; for others it can be a relatively simple setup. The key is the systems and processes that lie behind it rather than the facility and colocation itself. Either way, it is a highly recommended step for any operation, as long as it is approached in the right way.

· **Systemically learn and improve on the whole cycle:** This is the part of the process that should be about systemically learning from what works well and what doesn't, and it is what drives continuous refinement of the process. However, in the current absence of tight integrated planning processes, disciplined execution, and standardisation, it is impossible to leverage the full value from this systemic learning stage. In the current environment, we occasionally see attempts made to capture learnings in the form of a standard, but this usually ends up as a document that is never referred to, and within a few years no one even remembers that it exists.

This is why a fully integrated operating model, which centres around tightly managed planning, execution, and improvement, is so important. Without it, you are destined to forever run around in circles. It is why planning processes are currently usually adequate, at best. If planning is viewed as a discrete activity, it is easy to convince yourself that you have a good process. However, planning is not a discrete process. It relies on execution and improvement to make it effective. The other two are not separate processes; they are integrally linked. When you look at it this way, you start to see why the industry struggles with standardised, predictable, sustainable performance.

Currently, the mining companies who are best in the industry at integrated planning are doing parts of this process very well. But these examples typically centre around the logistics chain, where this is a bottleneck. As a result, dynamic management of the stock, transport, and quality-changes around the constrained logistics system is handled relatively well. However, these examples are still far from fully optimised, because they don't always look at the whole value chain, just the critical part of it. And they don't go deeply enough into the mining operations, so the mined stocks are viewed as the starting point for integration rather than the ore body itself. The variability and lack of predictability of the mining process, caused by the weaknesses already discussed, constrain the ability to truly optimise the logistics chain.

It is important to keep value front and centre in your thinking on all of this. All of your efforts need to be driven by value. But that's not a reason to stay out of the detail. It is essential that someone dives into the detail and makes sure the very complex value trade-offs are understood, so the business is always sure it is making the right judgements and can be confident that it is being run in the optimum way. Clearly, fully integrated planning is not a simple part-time role. It is something that warrants considerable effort. Although the role itself is a complex one when done right, a big part of its challenge is to bring clarity, alignment, and simplicity to the focus of effort across the business.

PRINCIPLES

It is worth highlighting two particular factors that often get in the way of developing good plans: the assumptions behind the plans and the psychology at play in setting these assumptions.

Assumptions

One of the biggest challenges (and weaknesses) in the planning process is the assumptions used. What is the right basis for the planned numbers?

- Is it nameplate capacity? This is the capacity as designed. How do you calculate this? Is it the original mine design numbers used in the original project? How do you factor in the many changes over the years?
- Is it last year's plan? Last year's actual? Previous best performance? Recent performance? Again, how do you calculate this? Is it based on the performance of the whole system, or the sum of the capacities of the individual parts of the process?
- Is it based on a whole-system simulation? Do you use some internally agreed standard for calculating 'demonstrated capacity'? Or is it a judgement call? How do you reconcile it back to previous plans? How do you reconcile it to planned improvement projects? Do you add in some extra as a 'stretch', or a 'target' beyond the budget? How much money and resources are you prepared to throw at making up any capability shortfall?
- What assumptions do you make for known unknowns like weather? What about unforeseen unknowns? What level of probability do you assign to the numbers (e.g. certain, prob-

able, possible, unlikely)? To what extent is it all a moot point because previous commitments to the press at a corporate level have already set the expectations?

The reality is, there are no hard and fast rules, and it is a judgement call for the business regarding how you approach this. The key is to ensure a common view on the expectations. This is where things start to break down quickly, because there is usually considerable misalignment between the different levels of the organisation on this. The problem is compounded when the final plan is decided through a combination of the individual plans rather than a consideration of the whole. The theory of constraints has proved that the capacity of the system as a whole is never as high as the sum of the parts, and the use of 'average' capacity information makes this a particularly dangerous exercise.

Psychology

Then there is the counterproductive psychology behind the planning process. Often, managers are 'coerced' by well-meaning bosses into accepting plans that they know are unrealistic. It is important to understand the mindsets that sit behind this, and the reasons why this can be a very bad idea.

For example, a manager is pulling together his annual plan and, after analysing the numbers, he approaches his boss. The conversation goes something like this:

- Manager: 'Boss, I know we did twenty units this year, but I've looked at the numbers for next year, and we are only going to be able to do eighteen. The reason is that we are in a particularly constrained area and we have some big maintenance shutdowns planned.'

- Boss: 'But I want twenty-two units next year'.
- Manager: 'I know. That's what I wanted too, but it's just not going to be possible. The constraints and shutdowns are just not going to allow it.'
- Boss: I don't think you're listening to me. I want twenty-two units. You go back and sharpen your pencil and tell me how you can make it happen. If you can't do it, just let me know, and I'll find someone who can.'

The manager goes away feeling chastened and looks at the numbers again. He feels bad that the availability figures haven't been as high as he wanted for the past couple of years, so he thinks there should be some upside in that. The same with the productivity figures. The shutdown may possibly be able to be shortened if everything goes well. And they are working on a few innovative ideas to improve the constraint issues, which hopefully will work out. So, he goes back to his boss.

- Manager: 'I've had a look at the numbers again, and it *is* theoretically possible to achieve the twenty-two units, but everything has to go right. We need to get our productivity levels up to where we really want them to be, the shutdown needs to go perfectly, and the trials on the constraint need to work out how we hoped. If all of that comes off, then the numbers work out at close to twenty-two units.'
- Boss: 'Okay, good, so we're agreed. The plan for next year is twenty-two units. But remember, this is *your* plan, so don't let us down.'

In reality, what just happened is that the manager walked out with a plan of twenty-two units for an operation that can likely only do eighteen. The shortfall of four units doesn't exist as far as the boss is concerned. The pressure then remains on the manager for the next twelve months to either just try their best and fail, or cut some corners and make some short-term decisions in an attempt to get close to the impossible plan. At the same time, there is likely a similar process happening on the cost budget. Meanwhile, on the front line, the operators look at the impossible plan figures, shrug their shoulders, and say, 'Here we go again', and they end up using it as just a general guideline, which they only broadly follow, with lots of degrees of freedom.

Sadly, this is a very common situation across the industry. What should happen is this (similar start to the conversation, but a very different end):

- Manager: 'Boss, I know we did twenty units this year, but I've looked at the numbers for next year and we are only going to be able to do eighteen. The reason is that we are in a particularly constrained area and we have some big maintenance shutdowns planned.'
- Boss: 'But I want twenty-two units next year'.
- Manager: 'I know. That's what I wanted too, but it's just not going to be possible. The constraints and shutdowns are not going to allow it.'
- Boss: 'Okay, so we have a shortfall of four units. Let's have a look at what we can do differently to make this up. You go away and see what you need to make up as much of this as possible, and I will discuss this with my colleagues to see whether we can make up the rest of the shortfall elsewhere.'

After some additional study, a combined plan is drawn up, and additional resourcing is agreed to. The difference from the first scenario is huge. This time the four-unit shortfall has not been swept under the carpet; the boss confronted it, found a solution, and applied appropriate additional resourcing. The manager has a sensible plan, which he feels committed to and realistically accountable for. And the frontline operators look at the plan and say, 'Finally, we have something realistic we can work to', so they follow the plan closely.

The first approach is nonsystemic. It relies on a personal management style, pretends the problem doesn't exist, sets the operation up for a year of pressure, and weakens accountability because no one in the manager's team expects to be able to hit the plan, let alone exceed it. The second is systemic, confronts the problem collaboratively, ensures the business has a realistic plan, and leaves the manager's team feeling motivated that they actually have a sensible plan for a change.

SOLUTION

The solution is obviously aligned with the second approach. The key points are: be realistic with your assumptions, confront any shortfalls to expectations, solve problems collaboratively from a whole-system perspective, provide additional resourcing where needed, and keep the focus on disciplined compliance to plan in the front line.

This process is considerably assisted if the 'planning and improvement' function is separated from the 'execution' function. And even more so if there is a separate 'capacity assurance' function within 'planning' which independently assesses operational performance and helps to decide what constitutes 'realistic' in the plan. If this is set up in the right way, within a well-structured integrated planning function, it can result in a breakthrough in terms of an operation's record of compliance to plan.

There are examples in the mining industry where years of shortfalls to plan have changed almost overnight to years of hitting plan, along with an associated massive change in the morale, mindset, and culture within the business. In one example, the simple step of establishing an integrated planning function, raising the profile of this role with a clear mandate, and providing greater transparency of information across the whole operating system increased the overall capacity of the operation by 10 percent, adding billions of dollars to the bottom line of the business.

CHECKLIST

- How are your plan numbers decided on?
- Is there a consistently applied process for aligning and deciding on the basis for the assumptions?
- Is this a collaborative process, which is approached from an integrated whole-system perspective?
- Who is accountable for this?
- Do you have a well-defined integrated planning function?
- What level of independence is there in the planning process, and especially in the analysis and decisions around the assumptions ('capacity assurance')?
- Is the 'fox in charge of the henhouse' with regard to setting the plan and executing it?
- Is 'improvement' integrally linked to your 'planning' function?

3.4 DISCIPLINED EXECUTION

OBSERVATIONS

If the plan is pulled together in the right way (well-informed, optimised, integrated, realistic), then the execution function should really just be about two things: disciplined execution of the plan and dynamic changes addressed in a whole-system-value context. Both of these sound easy but are generally poorly done in operations.

Hard-wiring the execution process

The starting point for this is to get very clear on what you are executing. Yes, this includes the overall headline production numbers from the planning process, but there is a lot of further detailed breakdown of this that needs to come from the execution teams themselves. If you want clarity and discipline in the way you execute, then you need to flesh out this detail. There are three components to this:

- Integrated scheduling
- The broader business management process
- Management of everyday dynamic changes

All three are necessary to achieve disciplined execution. This is discussed below, including the strengths and weaknesses of current general practice in the industry, as well as the requirements of a fully integrated system.

Integrated scheduling

Scheduling is essentially the process of turning the plan into a detailed sequence of work. It specifies the what, where, when, who, and how. In a mining context, it defines which areas will be mined, in what sequence, by which crew, and using what equipment. It also defines when shutdowns will take place and what maintenance work will be done.

Currently, in most mining operations, this involves at least three different teams coming together: a services team who prepare the ground, a production team who mine the product, and a maintenance team who maintain the equipment. The norm in the industry is for these three teams to produce separate schedules, and then a planning meeting is held to communicate these to each other. Any co-commitments and conflicts are jointly discussed, and then the three schedules are refined accordingly. This is all fine, as long as everything goes exactly according to plan. But it never does, because there are always unforeseen issues with people, equipment, timing, digability, etc. So, very quickly, things become misaligned between the teams, because each team only tracks their own schedule. The next day, the three teams get together to discuss what went wrong yesterday, and how each of their respective schedules have now changed. Efforts are then made to realign these, but it is a reactive process with constant surprises, and it requires continual changes to three separately managed schedules.

This is why *integrated* scheduling is so important, and represents a breakthrough methodology in terms of process efficiency. What should happen in a fully integrated system is that the three teams produce a single schedule that incorporates all three sets of activ-

Disciplined execution needs to be hard-wired into the day-to-day management process

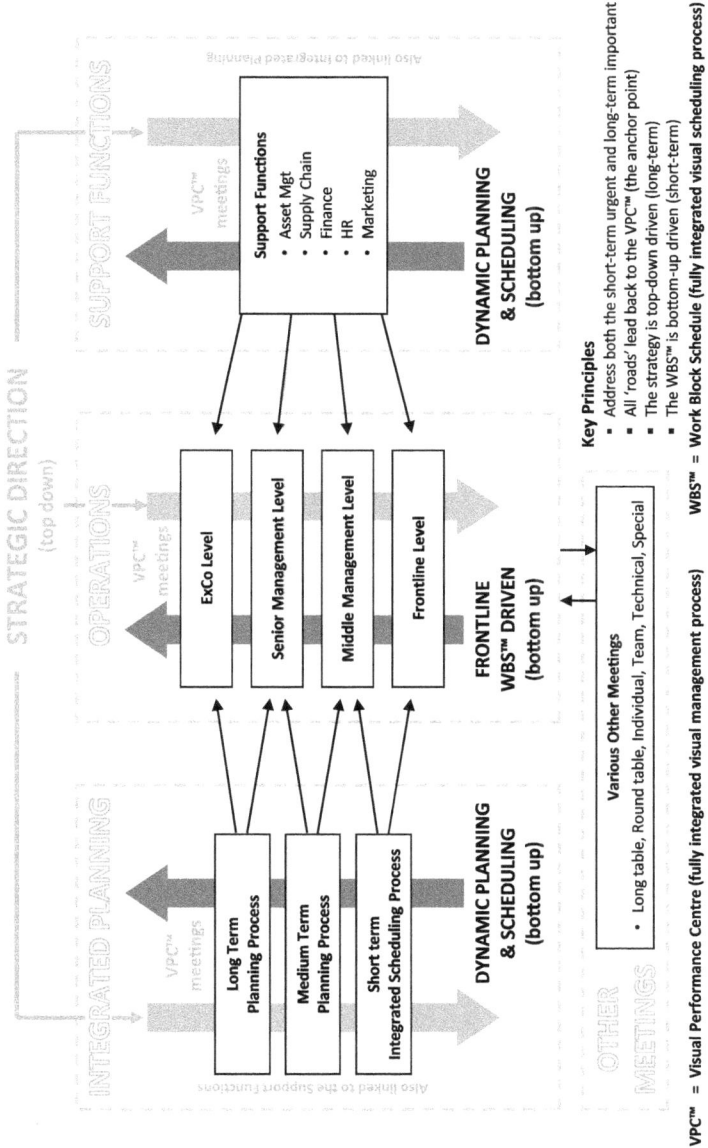

INTEGRATED PLANNING

Also linked to the Support Functions

VPC™ meetings

- Long Term Planning Process
- Medium Term Planning Process
- Short term Integrated Scheduling Process

DYNAMIC PLANNING & SCHEDULING (bottom up)

STRATEGIC DIRECTION (top down)

OPERATIONS

VPC™ meetings

- ExCo Level
- Senior Management Level
- Middle Management Level
- Frontline Level

FRONTLINE WBS™ DRIVEN (bottom up)

SUPPORT FUNCTIONS

Also linked to Integrated Planning

VPC™ meetings

Support Functions
- Asset Mgt
- Supply Chain
- Finance
- HR
- Marketing

DYNAMIC PLANNING & SCHEDULING (bottom up)

OTHER MEETINGS

Various Other Meetings
- Long table, Round table, Individual, Team, Technical, Special

Key Principles
- Address both the short-term urgent and long-term important
- All 'roads' lead back to the VPC™ (the anchor point)
- The strategy is top-down driven (long-term)
- The WBS™ is bottom-up driven (short-term)

VPC™ = Visual Performance Centre (fully integrated visual management process)   WBS™ = Work Block Schedule (fully integrated visual scheduling process)

Fig 3.4 Disciplined execution

ities, at least as far as it relates to the working interfaces between the teams. This needs to be made highly visual and forward-looking. Most importantly, there needs to be a strong discipline around compliance to the schedule. All three teams should be measured not on their own metrics but on compliance to this unifying schedule. The result of this is transformational. It completely changes the nature of the conversation between the teams.

With patched-together schedules, siloed behaviour is common: circular finger pointing between the teams ('production shortfall caused by breakdowns', and 'breakdowns caused by production not making equipment available for maintenance'), or indifference ('I'm okay; it's not my end of the boat that's sinking'), or siloed competition ('Their poor performance just makes me look better and gives me a chance to catch up where I was falling behind'). The conversations are all focussed on what happened yesterday and what the game plan is for today.

With a single *integrated* schedule, the behaviour is completely different. Suddenly each team is interested in each other's performance, because it is 'one fail, all fail'. The conversations automatically shift to being collaborative, forward-looking, and flexible to changing conditions: 'How can I help you? If you can shuffle the timing of that maintenance next Tuesday, I can give you extra time to get it done early, this Friday, by moving some of my equipment around'.

There are some specific ways of approaching integrated scheduling, which solve many of the practical obstacles that have held back progress in this space in the past. These are discussed in the next Principles section.

Broader business management process

Integrated scheduling deals with integration of the operation at a local level. However, there is also alignment with other parts of the operation to be considered, as well as management of other business priorities.

Currently, in many mining operations, these other activities are all managed separately, by different teams, through different meetings. This leads to a lot of misalignment, fragmentation of effort, and confusion. This tends to drive individual teams to step away from collaboration, and focus on delivering on their own priorities and metrics, as they view the alternative to just be too hard.

Instead, businesses need a mechanism for bringing all of these key activities together. Many have taken a big step forward in this regard by implementing a traditional 'Lean' visual management process (also known as Information Centre meetings, or SIC/short interval control meetings). What these generally have in common is that they are visual, short, and sharp, numbers-based, action-focussed, stand-up meetings, which bring together key people from multiple teams periodically (shift, day, week, month, as appropriate to the nature of the group). When properly led, this is a very effective way of breaking down siloed behaviour, but it usually doesn't go quite as far as the integrated scheduling approach of having a single goal that unites everyone.

Sadly, despite the considerable advantages of the 'Information Centre' type meetings, many operations still run the old-style morning meetings, which are long and drawn-out (one to two hours); and sit-down meetings, where people grab their coffee and get comfortable, and the chairperson goes around the table asking each of the long list of attendees to talk about how yesterday went and what their issues are for today and tomorrow. This leads to long discussions about issues and going off on tangents into problem-solving, which don't reach a conclusion or decision before time runs out. People are usually frustrated by the time they are wasting listening to debates about things that don't concern them. They regularly leave these meetings feeling frustrated by the lack of clarity around decisions and actions.

However, what is clear and consistent across the industry is that there is a very wide variety of formats and methods of application for these meetings, and relatively little standardisation even within businesses. Often, they are well-applied in one area but not in another, or well at one hierarchical level but not at the level above. Or they include gaps between, say, planning/execution/improvement, or long-/medium-/short-term functions. This immediately highlights the challenge of truly integrating activities and behaviour. This lack of standardisation is surprising, given the importance of these meetings to the way the business is run.

What should happen is that every business should recognise that the structure and process of these meetings is absolutely critical to business success. They represent the primary mechanism for defining and hard-wiring both the business strategy and operating philosophy. Collectively they should include everything that is important to the running of the business, both short and long term.

They should bring together the shorter-term, bottom-up operational production requirements and the longer-term, top-down business strategy requirements. There should be one meeting for each functional area across the whole value chain, including support functions, and these should be cascaded through every management layer of the business from front line up to the executive team. Each should have a carefully designed visual board, which provides structure and transparency to the information and ensures alignment of focus across the business. The boards and meeting process should be standardised across the business and entrenched as part of the 'way we work', so that when people move roles, the whole system continues to run seamlessly, because everyone is completely used to the methodology. The integrated scheduling process should feed into these integrated visual management meetings.

It sounds hard, and undoubtedly this is the primary reason why it has not become the norm. But is it really harder than the alternative, which is lack of alignment, frustration and friction, waste, rework, reinventing of the wheel, lack of stability and predictability of performance, and inefficiency? These are all associated with the current way of doing things, and ultimately lead to huge untapped latent value being lost in the interfaces between the parts of the organisation, and the way the business is run as a whole. It requires effort to design, manage, and sustain such a system. However, when you really think about the value involved, it is hard to defend why this isn't one of the top priorities of every business.

There are specific ways of approaching these integrated visual management meetings, which solve many practical obstacles that have held back progress in this space in the past. These are discussed in the next Principles section.

Management of everyday dynamic changes

If it was just a matter of planning and executing, life would be relatively easy. But this isn't the case. The reality is that dynamic changes (e.g. breakdowns, weather events, and other constraints) are constantly occurring in every part of the operation. These are usually hard to deal with, because by nature they are unforeseen, and they have a knock-on impact for other parts of the operation.

The production teams (executers) are badly affected, because they need to deal directly with the situation. The planners and schedulers are badly affected, because they now need to adjust their plans. Management is affected, because they now need to communicate the changes to other affected parts of the operation, and understand and deal with any further issues caused by this. For bigger issues, this is usually a frantic time, and the production teams have to drop everything to deal with the crisis. Whether it is a minor crisis or a major one, it is still a big distraction from normal operations.

There is often no real structured system or process for dealing with these changes. It is simply viewed as one of the core roles of the supervisors or managers involved to work out the best solution. That is fine, except it would still make sense to provide some structure, process, and support for these leaders. In the absence of this, they simply have to drop everything and deal with these situations. It is not uncommon for frontline leaders to have their entire day consumed by such issues. This includes trying to understand what has happened, getting to the bottom of the truth, finding the right people to solve the issue, tracking down parts, negotiating priorities, contacting other affected parties, informing their bosses, obtaining approvals, completing paperwork, etc.

There are two ways of handling this challenge. First, you can just plan in a less rigorous way, which keeps you out of this detail. However, this allows degrees of freedom to creep in with regard to how activities are performed, which pulls you away from the standardised, predictable approach you are seeking. Or, second, you can embrace the detail and manage the dynamic changes in a systemic way. The detail is where most of the conflicts and misalignment occur, so this has major advantages.

The norm in the industry is the former, i.e. to stay out of the detail. And this is a key weakness with currently accepted operating models. Without the detail, you can't hope to align all of the parts and achieve disciplined execution. Without disciplined execution, there is no point in putting a lot of effort into the planning. So, planning and execution typically get managed with relatively high-level figures, with the justification that it provides flexibility to the planning and execution teams. Which would be fine if it worked. But it doesn't work effectively. There is generally a great degree of misalignment and frustration between teams within the workplace.

Hence the importance of the integrated scheduling process, which provides this detail for the production front line. Equally, the importance of the integrated visual management meetings, which provide this detail for the management of the broader business requirements, including a link to the integrated scheduling. Both of these meetings (and their associated visual boards) should be specifically designed to help deal with dynamic changes. They should not just be a reflection of the plan and actual performance against the plan. They should very much be live boards that capture and confront changes and variances to the plan, with a view to constantly ensuring everyone is on the same page with regard to how to deal with these situations. And it's important to make sure that actions are being taken to get back online so that longer-term business expectations and commitments are met.

The integrated scheduling process, and the transparency and collaborative behaviour it drives, is a major advantage in dealing with these situations. The integrated visual management meetings are also a big help in ensuring bigger issues are handled in the best way for whole-system value. Several other aspects of the overall holistically design-integrated operating model also assist.

Some operations have taken this to another level by introducing an Integrated Operations Centre (IOC), which is a highly structured way of managing these dynamic changes. This includes a 24/7 team whose job it is to track and manage the dynamic aspects of optimising scheduling across the whole operating system. For some operations this is a complex facility; for others it can be a relatively simple setup. The key is the systems and processes that lie behind it rather than the facility and colocation itself. Either way, it is a highly recommended step for any operation, as long as it is approached in the right way.

PRINCIPLES
What about 'Lean'?

There is a distinction between 'Lean' as applied in a business like Toyota (where the improvement culture is highly advanced and most of the basics have become embedded almost subconsciously) and how it is applied in the mining industry. 'Lean' has been tried by many mining companies, but for the most part it has not stuck particularly well. Despite some great benefits delivered successfully (e.g. the 'Lean' Information Centre boards), few have stood the test of time, and there are examples of rigorously implemented 'Lean' systems being embedded and producing good results for companies, but ten years later they end up as a shadow of their former self.

This is not a reflection on the value of 'Lean'. It is a reflection on the relatively low cultural maturity of the mining industry, with respect to continuous, systemic, sustainable improvement. Or, rather, the extremely high cultural maturity within the manufacturing industry. It is clear that the mining industry has some serious structural weaknesses when it comes to improvement culture. Businesses are not getting the operational basics right, and the systems and processes used are not set up in a way that is systemic and sustainable. So, in all reasonableness, it is not surprising that 'Lean' hasn't yet found a solid home in the mining industry. Other industries have had mixed results, depending on whether their improvement culture is closer to Toyota's or the mining industry.

The cultural foundation that makes 'Lean' work in Toyota, and which they take for granted, will be provided by the fully integrated operating model proposed in this book. Many of the principles of integrated operations are just embedded in culture and taken for granted at Toyota and have been forgotten as a formal process. One way of looking at the new fully integrated operating model is that it will provide this platform.

How to take integrated scheduling and integrated visual management meetings to the next level

The key elements discussed in the previous Observations section illustrate what integrated scheduling and integrated visual management meetings should ideally look like. These features have been successfully combined into two highly complementary processes in NextGenOpX's Work Block Schedule (WBS™) and Visual Performance Centre (VPC™) designs. These have taken the best of existing systems and processes across the industry and enhanced them using the principles of integration, as well as customising them to suit the practical challenges presented by the mining industry. This combination has led to a real breakthrough in the quality of design, as well as the rigour of application and sustainability of outcomes.

Let us take a moment to discuss the difference between the traditional 'Lean-type' meeting, and the fully integrated WBS and VPC approach. These both would apply in any industry, but I will discuss this in the context of what has been successfully implemented in mining applications. In other industries, the visuals would be different, in the same way as they differ across different types of mining, processing, and logistics operations.

What makes the WBS and VPC approach different is the way it is practically designed to suit an operation that does not have the same high degree of maturity in improvement culture as a manufacturing plant does. The approach is much more visual, not just in numbers and graphs, but particularly in the visual and spacial context provided, which is such a feature of mining. These visuals are also designed to be informative and educational for the workforce. This is important in a business where tenure and experience are often lacking, and the geographical spread of the operations often means workers are ill-informed about other parts of the operation.

An important distinction of the WBS is the focus on the single, unifying, integrated schedule rather than coordinating separate schedules. It considers a 'block' of work, which involves several different teams coming together to produce a collective result. That's why the term *Work Block Schedule* is appropriate terminology to use. The WBS can be applied to various levels of complexity. It targets the interfaces effectively, and the focus on compliance to schedule helps to tighten the degrees of freedom that are the norm in current operations. The granularity provided by the connected daily, two-week, and three-month forward-looking schedules used in the WBS process are also a significant improvement. This helps to deal proactively with variance management and achieve the desired outcome of predictability.

The level of rigour in establishing these WBS and VPC boards right across and down through the operation is also a distinguishing feature. Such boards/meetings are often viewed as a middle-management requirement, which does not always go right down through to the front line or right up to the executive committee. And where this view is taken, it is a mistake and a weakness, because this is part of the reason why fully integrated planning, disciplined execution. and sustainable improvement are never achieved.

One of the biggest surprises to teams when they design the content for these boards is that they are embarrassed to discover how unclear they really are about their own part of the operation/

business. Teams quickly discover that some basic questions are very difficult to answer, such as, 'What are our most important short-term priorities? What are our most important long-term priorities? What behaviour are we trying to drive? What are the best metrics for this? What are the root causes of our problems? What actions are we taking to fix these? Who is accountable for this? Who is tracking it? How do we govern all of this to make sure it doesn't just get lost in all the confusion?' They are so used to just dealing with mountains of information and data that they never have the chance to actually sit down and think about these important questions.

It is hard work to put a system like this into place, as well as to clarify in detail the role of the multitude of teams across and down through the business along with standardising the format. But, again, you have to ask the question, 'What is the alternative? Just allowing everyone to figure out this complex puzzle for themselves?' Surely not.

The value attached to this is enormous. The misalignment and inefficiency across the interfaces, the impact of unsustained improvements, and the cost of constantly reinventing the wheel all cost businesses dearly. It is absolutely worth taking the time to set it up properly.

Board designs

Work Block Schedule (WBS™): There is an art to the way the integrated schedule board needs to be drawn up. It should comprise a daily (broken down into shifts), two-week (broken down into days), and three-month (broken down into weeks) schedule.

These must be interlinked by a rigorous management process (by shift, daily, weekly, monthly...with the weekly and monthly sessions looking increasingly broad, and bringing together other sections and functions that play a part in the overall performance of the operating system). This must be underpinned by careful training in how to lead and participate in meetings. The format must be highly visual, interactive, and in the form of an interactive board. The layout of the boards should be codesigned with the people involved.

Visual Performance Centre (VPC™): Similarly, there is an art to the way the VPC board needs to be drawn up. It should comprise a short-term section and a long-term section. The short-term section (production driven) comprises: visual context; safety, health, and environment; people issues; equipment issues; and the WBS itself. The long-term section (strategy driven) comprises: performance trends, reconciliation, projects, and risks. And lastly, there is a section for actions and celebrating success.

Digital: These boards lend themselves very well to digitisation, and it simplifies the whole process of managing and updating information. There are good examples within the industry of digitised VPC boards. One potential downside of *digital* boards is that you generally use one board for multiple screen views, and the board can be switched off, which detracts from the transparency of displayed information. You also lose the benefit of tactile accountability for the manually entered information when meeting participants update their own data just prior to the meeting. On the other hand, a fully digitised board is much simpler for upkeep, and it has the benefit of being easily available electronically to everyone. However, a high-resolution photo of a manual board can also be easily copied to all, so don't overplay this latter point. In summary, this should be a matter of personal choice for the business. It's best handled as a transition from manual to digital, rather than starting off digital if the teams are not yet used to the process.

Meeting design

The design of the architecture of meetings across the business is just as important as the board designs.

In most operations today, there are a multitude of meetings. In fact, meetings are held in order to prepare for meetings. It is common to hear people say, 'I've been busy all day in meetings'. It is worth reflecting on the meaning of 'busy' in this context. 'Busy' is very different than 'productive'. When people (particularly leaders) are stuck in meetings all day, it means they aren't interacting with their teams, staying in touch with their operations, and dealing with issues.

All meetings should have a purpose. And most importantly, they should all feed into and off the VPC process. If they don't have relevance to the VPC content (whose purpose is to reflect everything that is important to the business), either as an input or an output, it is worth questioning the value of the meeting.

The key here is the VPC meetings should lie at the heart of a broader architecture of meetings. There are various types of meetings: individual, team, corporate/head-office, technical, special, and many others. Given the amount of time that large groups of key people spend sitting in a room, often in an inefficient and ineffective meeting process, it is worth carrying out a careful review of all of your meetings to understand and map out the overall architecture of meetings across the business. This should include the following: the purpose, the agenda, the required attendees (additional hangers-on should be strongly discouraged), and how each meeting ties in with the VPC process.

This is also one of the key elements of an overall decluttering and simplification process that should be carried out as part of the design and implementation of a fully integrated operating model. A 50 percent reduction in time spent in meetings is a sensible goal for most operations, and a number have been able to achieve that through a rigorous review. The positive impact on morale of getting leaders to spend more time in the workplace is a tangible benefit.

SOLUTION

Overall, the solution requires the following:

- Get really clear on your business strategy, your operating philosophy, and the way the parts of the business all need to work together to achieve the desired end result.
- Hard-wire this into a well-structured combination of integrated scheduling (WBS™), visual management boards (VPC™), and meetings.
- Carefully align these across and down through the business (across all parts of the operating system and support functions, and down through all management levels).
- Review, simplify, structure. and align all meetings, and ensure everything feeds off or into the VPC process.

- Train your people in the proper running of, and participation in, meetings.
- Establish a mechanism for helping with the management of dynamic changes.
- Give strong consideration to establishing an Integrated Operations Centre (IOC), designed fit-for-purpose to your needs.

CHECKLIST

- How is your frontline scheduling carried out? Is it pieced together with siloed schedules, or a single, fully integrated schedule?
- What is your broader business management process?
- What is the quality of your meetings? Who attends? Do they cover all of the short-term production priorities? Do they link to the plan/schedule? Do they monitor compliance to plan/schedule? Do they cover all of the long-term strategic priorities? What information is discussed? Are actions tracked and followed up? How visual are they? What is the general tone of the meetings?
- Do you have a picture of all the meetings currently taking place?
- How much time do your leaders spend in meetings, as opposed to out in the field?
- How are dynamic changes managed?
- Do you see value in a fit-for-purpose Integrated Operations Centre (IOC) for your operation?

3.5 CONTINUOUS, SUSTAINABLE IMPROVEMENT

OBSERVATIONS

There is a big difference between driving improvement and driving *sustainable* improvement. It is worth reiterating the mantra 'An improvement isn't an improvement unless it is systemic and sustainable'.

Short-term improvement just needs focus, resourcing, and a 'big stick' approach. In fact, sometimes just the Hawthorne effect achieves results (this is when people modify their behaviour and things improve, just because they know they are being observed). Sustainable improvement means designing a working environment and culture, such that when the 'big stick' goes away, the improvement continues.

The 'big stick' approach is unfortunately how a great deal of improvement is achieved in the mining industry, and this is why productivity has declined, and why the same old initiatives resurface every few years—cost cutting, head count cuts, inventory reductions, organisation structure changes, etc. Each time, these initiatives are presented as bold new steps to reposition the business. They are not. They are the same old, tired steps that haven't worked sustainably in the past.

For those in the workforce who have been at an operation for decades, it is depressing to have to go through these cycles time after time, knowing that the results will be the same: some short-term gains (usually cost cutting, because it is easiest), followed by years of turmoil, while they have to deal with the consequences of poor, short-term decisions. These are gradually followed by all the same things creeping back in again. And repeat!

The industry is crying out for a better way than this. A smarter way. A way that invests more in positioning businesses for the long term than in simple, quick fixes.

It is time for business to step up their game. It is time for the boards and senior executives who run mining companies and set the direction and expectations of their businesses to invest more effort into making it easier for their teams to succeed. As already discussed in the previous chapter, there are some structural weaknesses in how mining operations are set up today. The reason these are not already central to the current agenda may be that the 'better way' has not been readily apparent to date. And, for the most part, everyone is as bad as everyone else in this regard, so there is no perceived burning platform.

Well, the burning platform has arrived. It is the culmination of a number of things: the acceleration of new technologies, the danger of industry disruption, the increasing complexity of running operations, the declining 'capability' of management teams, changing expectations of the younger workforce, and frustrations with inefficient and cumbersome systems in the new world of instant gratification, to name a few.

There *is* a better way. The key difference is the focus on systemic and sustainable, rather than short-term, fixes.

Long-term performance doesn't have to preclude short-term performance. It just means you have to do things in a way that tackles

Focussing on systemic solutions, which address the cause of problems, not the symptoms

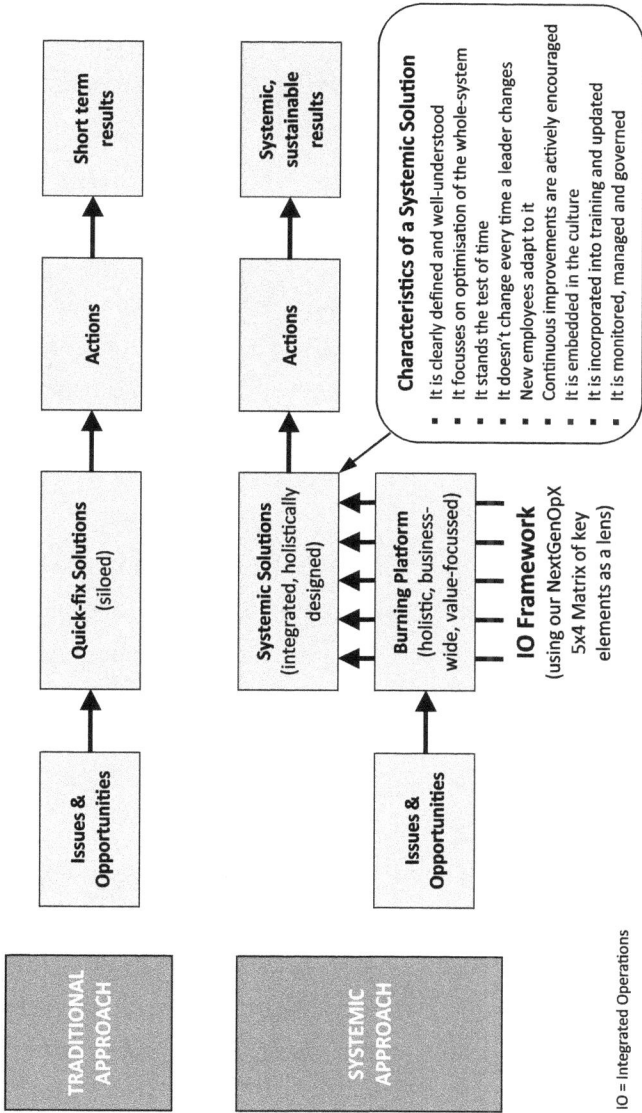

TRADITIONAL APPROACH

Issues & Opportunities → Quick-fix Solutions (siloed) → Actions → Short term results

SYSTEMIC APPROACH

Issues & Opportunities → Systemic Solutions (integrated, holistically designed) → Actions → Systemic, sustainable results

Burning Platform (holistic, business-wide, value-focussed)

IO Framework (using our NextGenOpX 5x4 Matrix of key elements as a lens)

Characteristics of a Systemic Solution

- It is clearly defined and well-understood
- It focusses on optimisation of the whole-system
- It stands the test of time
- It doesn't change every time a leader changes
- New employees adapt to it
- Continuous improvements are actively encouraged
- It is embedded in the culture
- It is incorporated into training and updated
- It is monitored, managed and governed

IO = Integrated Operations

Fig 3.5 Continuous, sustainable improvement

the issues properly so you achieve both. It is harder to set up in the short term, because you need to do some things differently from the ways of the past, but it all makes absolute common sense and is much easier for everyone in the long term. Not to mention the additional value it liberates and the platform it creates for future technologies and business growth.

PRINCIPLES

A systemic, sustainable approach is one that stands the test of time. It doesn't change each time a leader changes. It is standardised. It is embedded into every system and process within the business. It becomes a core part of the culture of the business. New employees don't change it; they adapt to it.

It requires a new type of behaviour: a focus on optimisation of the whole system rather than just its parts. Siloed behaviour is much easier, because you don't need to worry about anyone else. Systemic, standardised, whole-system decisions are harder, so the tendency will always be to default back to siloed and individually convenient decisions.

To prevent this, you need to embed the right behaviour into the DNA of the business: into the meeting processes, reporting, communications, incentives, roles, accountabilities, training, etc. If you just address one of these aspects and leave the others misaligned, they will confuse the message and gradually pull people back to the old way.

Standardisation: Having a standardised way of doing things, which everyone aligns with and commits to, is the starting point.

This needs to be codesigned and continually improved by the users. It needs to be visual and flexible enough to deal with different circumstances. 'Standardised' is not the same as a standard. The former is an outcome; the latter is a piece of paper. Your true 'standard' is what you *do*, not what you *say*. The documented standards are there as a training and auditing reference point to ensure you don't drift off over time. They should not be a tool to cover the backside of leadership who tolerate deviations every day, until something goes wrong and they need to find a scapegoat.

Standardisation should be a culture, something aspired to by the workforce because it makes their lives easier, better, and safer. It should be something they willingly play a part in achieving and continuously improving, because they understand and appreciate the difference it makes to the overall well-being of themselves, their colleagues, and the business as a whole.

Once you have stability in the improvements you make, you have the foundation for making a second performance lift through further optimisation, new technologies, automation, and digitisation. Without a stable operating platform, you will never achieve the full benefits of a progressive technology strategy and the additional value potential this brings.

Without standardisation you will always leave value on the table, because improvements can't be rapidly shared and leveraged across the business. This is particularly the case across a global business, but it also applies within a local business unit, and even down to a shift level.

Governance: The last important factor underpinning sustainability is governance. This is crucial because over time, systems and processes tend to evolve, both positively and negatively. Left unchecked, hard-fought gains will be steadily lost over time, particularly when a new paradigm in managing operations is involved. Positive evolution can be regarded as continuous improvement, which is important to encourage. However, this needs to be managed and enabled in a coordinated and controlled way. Negative evolution is when systems and processes gravitate back to the old way or the easy way. Again, this needs to be monitored and controlled in a structured and formal way. This should be a multidimensional, holistic approach, which considers operational, technical, organisational, and motivational factors.

SOLUTION

The right framework for setting up this new way, and the vehicle for driving the change, is integration.

It sets the right contextual objective in focussing on the whole system, and particularly the interfaces between all the multidimensional parts. It provides the holistic framework that allows the right behaviour to be embedded into every aspect of the business. It creates a 'pull' to standardisation, and it defines what governance needs to cover in order to be effective.

CHECKLIST

- If you look back over twenty years, how sustainable have your improvement efforts been?
- How well does your current technology programme consider the maturity of the operating platform it will sit on?

- How well-defined is your operating model, and how well-standardised are your operating practices?
- How do you govern this to ensure compliance and sustainability?

3.6 HARD-WIRING SAFETY AND RISK INTO YOUR ROUTINE MANAGEMENT PROCESSES

OBSERVATIONS
Safety

Safety has advanced a lot in the mining industry, since moving from a physical-conditions approach to a behavioural-cultural approach. Not all companies have cracked the latter, but those who did saw a dramatic improvement in injury rates.

Focussing on conditions (notably safety housekeeping), certainly improves safety, partly due to the removal of unsafe situations, but also in no small part due to the behavioural/cultural aspects that it drives. It is certainly a very tangible thing that can be measured and assessed. As such, it is a logical starting point for any safety programme.

Coming at safety from the other end, i.e. behaviour and culture, is much harder, but the rewards are much greater. Some of the best safety results achieved in operations have been at sites with relatively mediocre housekeeping conditions. The objective of the behavioural approach is to achieve a mindset in every employee that ensures they make the right safe decision to follow the safety standards to the letter, even when no one else is around to check on them. This is the ultimate goal in safety, but it also presents many challenges.

It requires training in the standards, and refresher training when there are changes, ongoing safety interactions to monitor and correct behaviours, and regular audits to ensure the standards are being followed. In the context of the sheer number of stan-

Safety and risk mitigation benefit from a well-designed integrated operating model

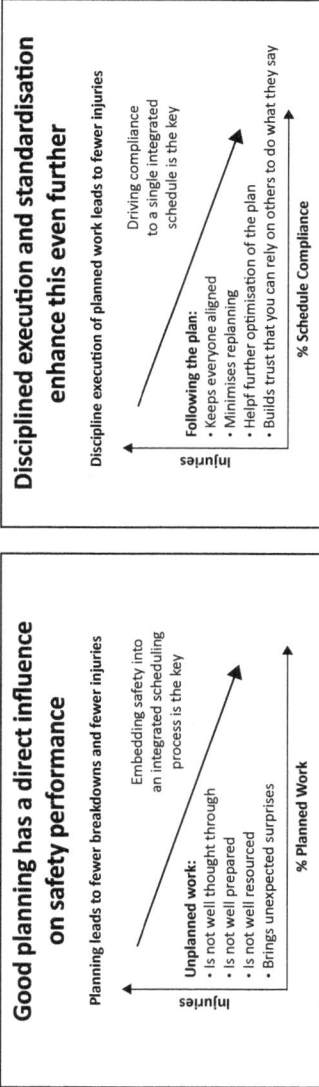

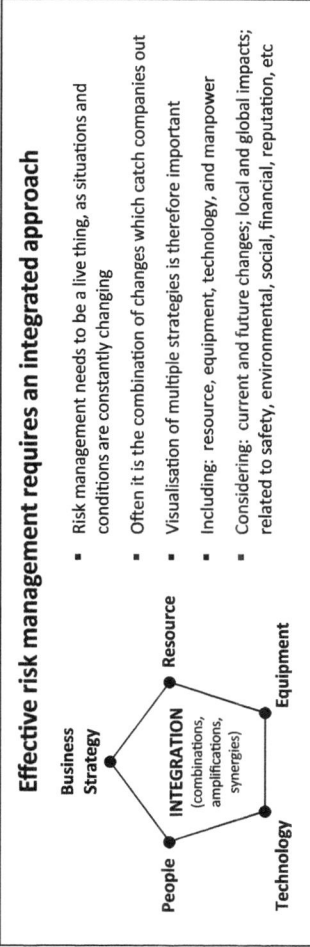

Good planning has a direct influence on safety performance

Planning leads to fewer breakdowns and fewer injuries

Embedding safety into an integrated scheduling process is the key

Unplanned work:
- Is not well thought through
- Is not well prepared
- Is not well resourced
- Brings unexpected surprises

Injuries (vertical axis)

% Planned Work (horizontal axis)

Disciplined execution and standardisation enhance this even further

Discipline execution of planned work leads to fewer injuries

Driving compliance to a single integrated schedule is the key

Following the plan:
- Keeps everyone aligned
- Minimises replanning
- Helpf further optimisation of the plan
- Builds trust that you can rely on others to do what they say

Injuries (vertical axis)

% Schedule Compliance (horizontal axis)

Effective risk management requires an integrated approach

- Risk management needs to be a live thing, as situations and conditions are constantly changing
- Often it is the combination of changes which catch companies out
- Visualisation of multiple strategies is therefore important
- Including: resource, equipment, technology, and manpower
- Considering: current and future changes; local and global impacts; related to safety, environmental, social, financial, reputation, etc

Business Strategy

Resource

INTEGRATION
(combinations, amplifications, synergies)

Equipment

People

Technology

Fig 3.6 Hard-wiring safety and risk into your routine management processes

dards and the relatively clumsy training methods, combined with regular employee turnover and constant management changes, this is no mean feat. Even those who do it well often struggle to maintain that standard.

What is interesting is that there has been relatively little effort to embed safety into the day-to-day management processes. A safety-share at the front of every meeting is certainly an excellent way of reminding everyone that safety is important, but that is not *embedding* safety into the management process; it is *adding* it onto the management process. This is not the same thing at all. After the safety-share, teams too often immediately move on to discussing the prior day and the day ahead, and they have come to accept that this is a rather chaotic and reactive space. In many operations, the safety and operational discussions are somewhat separated. Chaotic, reactive, and separate are not words that go well with safety.

As already discussed, plans and schedules are typically not well-integrated, realistic, dynamically managed, or strictly followed. Execution of the plans and schedules comes with many degrees of freedom and involves people from different teams making siloed decisions that are often not well-communicated to other parties.

In such an environment, culture can only take you so far on safety. Therefore, there is a big opportunity for companies to take the next step in improving safety by embedding it into the day-to-day management processes and decision-making.

A fully integrated operating model tangibly addresses current weaknesses in integrated planning, disciplined execution, dynamic decision-making, collaborative behaviour, and standardisation. The safety benefits are obvious.

By focussing teams on compliance to a single, integrated schedule, which is highly visual, easy to understand, and creates a 'pull' towards standardisation, it makes it much easier for people to do the right thing, in the right way, at the right time. The systemic, visual process helps to identify potential safety issues proactively, days ahead of time, which allows proper safety mitigation plans to be put in place. This makes safety proactive, disciplined, and collaborative.

These are all things that would benefit any industry.

Risk Management

Short- to medium-term safety risks are one thing. Medium- to long-term strategic business risks (and opportunities) are another. The problem with the latter is that they are too often treated as an annual risk review process. Because the risks are longer term, and any negative consequences are often viewed as an unforeseen surprise, they usually don't receive the same level of attention as safety. This certainly applies to the way risk is embedded into the ongoing routine-management processes.

Similar to safety, it is important that management of strategic risks (and opportunities) is tangibly woven into the ongoing management processes and not left to periodic compliance audits. There are obvious differences in where and how these are embedded, but the principles are the same. The point with risk management is that it is often impacted by longer-term strategic shifts. This warrants closer attention to the integration of four interrelated strategies, namely the resource, equipment, technology, and manpower. This can be dramatically improved through better visualisation of the strategies and interdependent impacts across these four areas. NextGenOpX has developed a good blueprint for this.

These represent the key variables which need to be strategically considered and managed over the short, medium and long term. They should consider various forms of impact: obviously direct financial impacts, but also both local and global reputational impacts of a social and environmental nature.

PRINCIPLES

Any good safety programme starts with conditions, because it is the easiest place to start—housekeeping, guarding, lighting, dust, noise, etc.

Focussing on behaviour and culture is typically the next step—'zero goal' philosophy, safety as a priority, standards, transparency of performance, increased weighting in incentive schemes, audits, safety interactions, incident reporting and investigation, injury management, etc.

The next step after this is to come back to the work processes: ensuring that work is well-planned and executed according to that plan, that dynamic changes are well-managed and well-communicated to those affected, that work is carried out according to standardised methods, and that the systems and processes behind this are simple and clear to all. These things are not generally done well in the industry. Not out of lack of effort, but because it is extremely difficult to achieve, especially when the operating fundamentals are not there to support it.

The sophistication in approach to business risk management usually goes hand in hand with the evolution of safety management, and it is often managed in parallel, by the same function. However, the nature of the challenges of risk management requires a different approach. Some aspects of risk management are fairly black and white, like compliance audits. However, it is often the longer-term strategic impacts that catch companies out when combinations of changes result in some major impact. So, the risk space needs to be more about the combination of issues rather than management of the individual issues themselves.

SOLUTION

An integrated operating model provides practical solutions that assist both safety and risk management, and as such, this represents a largely untapped opportunity for safety and risk improvement in operations.

Establishing an integrated scheduling approach, like the Work Block Schedule methodology, is a logical starting point for companies ready to move to the next level. This provides a rigorous mechanism for ensuring that identification and mitigation of safety issues is explicitly hardwired into the design of this.

As for risk management, the starting point for an already well-established platform should be to design a visualisation of the resource, equipment, technology, and manpower strategies, and build a meeting process around this that targets the interfaces, interdependencies, and synergies between these.

CHECKLIST

- How effective are your processes for integrated planning, disciplined execution, dynamic decision-making, collaborative behaviour, and standardisation?
- Which of these aspects are the cause of most of your injuries and incidents?
- What are you doing to address any current weaknesses?
- Do you ever sit down as a team and look at the interdependencies between resource, equipment, technology, and manpower strategies over the long term (life of mine) for your operations?
- Are your strategies currently even in a form that makes this possible?

3.7 LEVERAGING THE VALUE OF DATA

OBSERVATIONS

Data: Good data is an important starting point for any business. However, even in this technologically advanced age, it is still common to come across operations that have a problem with good data. Most companies with this problem recognise the importance and are in the process of rectifying this. Usually, businesses are doing this through a combination of better data-gathering systems, new sensors, digitisation, and data-integration platforms.

Information: Once the lack of data is solved, the problem quickly moves to realising that the issue is actually the need for information from that data. This is surprisingly difficult in a world where digitisation means there is an explosion of data available in micro-detailed form. Things like units, variability, averages, trends, and predictability all come into play. It depends on whether the information is used for after-the-fact reporting, ongoing local operational control, or dynamic whole-system decision-making. So, this is an area that most operations tackle, but not particularly well, and certainly there is plenty of confusion left on the table.

Decisions: Once you get good information from the data, the next problem is communicating it in a way that leads to good decisions. Again, this is much easier said than done, especially when it comes to cross-functional interactions. Just because you tell someone something important doesn't mean they will understand it, believe it, or respond to it. They are working in a different context, have different KPI priorities, and take their instructions from a different person. So, they are likely to view it as a judgement call whether or not to respond to it. This is a common issue when it comes to remote condition monitoring and operators are advised that a piece of equipment is heading for failure and needs to be shut down.

Getting this right requires clear decision rights, protocols, and mandates. It's not just the information per se. This is considerably complicated by the fact that most companies run the different parts of their operations in relative silos with no clear accountability (at least on a live, dynamic basis) for whole-system decision-making. Even when you have these decision rights and protocols in place, such as through a comprehensively mandated Integrated Operations Centre (IOC), this is not always a straightforward conversation. In most operations, converting information into good decisions is generally not handled well.

Sustainability: Once you have a solid process for analysing the data, and extracting and communicating the right information to the right people for good decision-making, you are in a good space...until someone critical leaves, and the rigour behind your good setup starts to erode. Too often, when operations are run well, this is because a few good people choose to do the right thing. When they leave, the whole system can come crashing down, because it relies on *them*, not a systematised approach.

The actual goal needs to be the setting up of a systemic approach, which stands the test of time and ensures sustainability of the data management, analysis, and decision-making system. This requires a holistic approach, which is integrated with the various aspects that make up a fully integrated operating model.

We seldom leverage anywhere near the full value of data analytics, software and 'digital'

The 'problem' is viewed too simplistically:

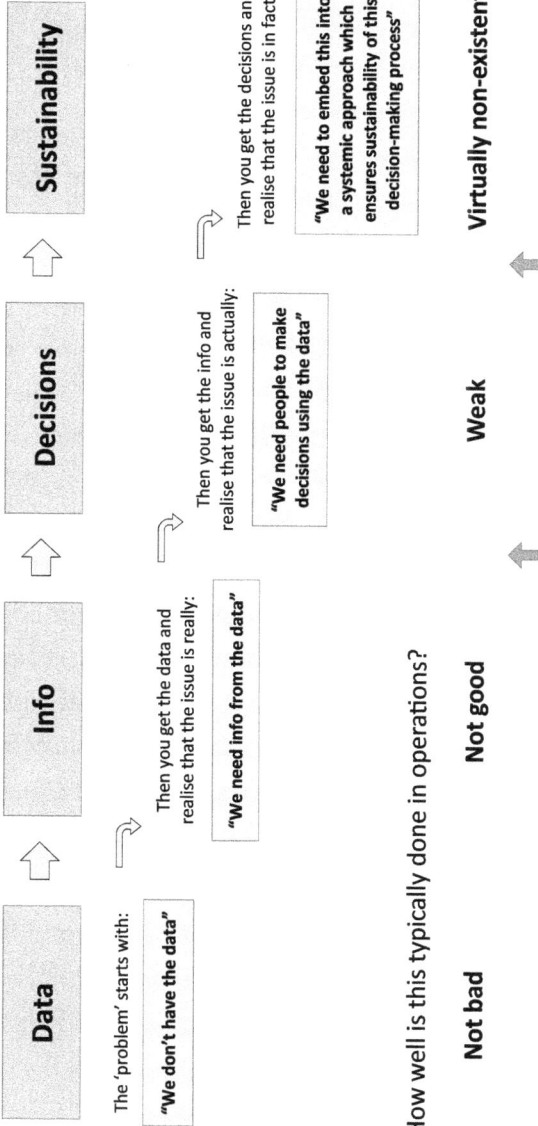

| Data | → | Info | → | Decisions | → | Sustainability |

The 'problem' starts with:

"We don't have the data"

Then you get the data and realise that the issue is really:

"We need info from the data"

Then you get the info and realise that the issue is actually:

"We need people to make decisions using the data"

Then you get the decisions and realise that the issue is in fact:

"We need to embed this into a systemic approach which ensures sustainability of this decision-making process"

How well is this typically done in operations?

| **Not bad** | **Not good** | **Weak** | **Virtually non-existent** |

These are the two interfaces which are the biggest challenge

Fig 3.7 Leveraging the value of data

PRINCIPLES

The following aspects are useful to consider in the data analytics space. They certainly apply to the mining environment, but most will also have application in other industries.

Purpose: Data comes in many forms. The way it is approached depends on whether the information is used for after-the-fact reporting, ongoing local operational control, or dynamic whole-system decision-making.

Value: There is not much point in having data if you are not able to interpret the information it contains. And there is not much point in interpreting the information if no one acts on it to add value.

Units: What is the best way of looking at the data? For example, for a live production number, is it tons or tons per hour? Or is it tons per shift or tons per day? If it's the latter, do you extrapolate the instantaneous reading to a tons-per-shift or tons-per-day equivalent? And if so, do you allow for the fact that there is generally a phasedown in productivity around shift change? When discussions are held about production rates, what unit is used, and does it differ depending on who you are talking to? Tons per shift for the operators, tons per day for the middle management, tons per month for the senior management, tons per year for the executives? Is it any surprise there is often a lack of clarity when everyone is talking different numbers?

Averages: One of the dangers of data analysis is that people often focus on averages, which can be extremely misleading, especially in a batch-type operation like mining. The average may be below the bottleneck rating, but if the average consists of high and low readings, it may mean that for part of the time the operation is constrained.

Variability: Variability is important to understand. It is the degree of variation around the numbers. That can be within a shift and affected by operating cycles. Or between shifts and affected by the experience and style of the different operators. Or between days and affected by maintenance cycles. Or between months and affected by weather. Whatever the case, it is important to understand these differences, because they highlight areas of opportunity for improvement.

Predictability: Predictability is crucial for truly optimising an operation. When you run in a reactive mode, you are always scrambling to recover. When you know what is coming, you can either take action to prevent it, or you can ensure everyone is ready to respond when it happens. The most advanced forms of analytics today are predictive analytics. They can interpret empirical trends, relationships. and impacts across multiple variables (including weather, time of day, time of week, holidays, crews, etc), and provide much better predictions of performance than just mathematical averages. As artificial intelligence (AI) becomes more prevalent, this will progress to a whole new level.

Trends: Trends are a useful way of looking at data, but are more difficult to communicate simply to operators in the field, unless you are in front of a computer. However, they bring an additional dimension to data, beyond just the absolute value, and they need to form part of the analytical process, in particular the value that can be derived from predictability.

SOLUTION

Establishing a fully integrated operating model is the right way of approaching this...with a clear operating philosophy and single game plan, where everyone understands the overall goal and their role within it, and decision rights and mandates are clear.

From this foundation, you can then define the KPIs, the escalation trigger points, and the right data and information formats to track, in order to manage the overall operating system in the optimal way. If you just jump straight into data as a specific element, you can quickly add to the confusion rather than improve it.

Any operation of significant size should have some sort of data analytics function. This is usually fairly obvious in the areas of, for example, mine dispatch (fleet management), and plant control and condition monitoring (asset management), although it is not always done well. However, the area that is very underdone is the analysis of whole-system performance, and particularly the connection to dynamic whole-system decision-making.

This is an essential requirement of any advanced operating model. It is significantly enabled and empowered by an Integrated Operations Centre (IOC), but this doesn't have to be a complex undertaking to start adding value.

CHECKLIST

- Are you stuck in the data and information space, or is your data analytics linked directly to decision-making?
- Do you distinguish between analytics for reporting purposes and for dynamic decision-making purposes?

- Is your data analytics focussed on whole-system optimisation?
- Do you analyse and focus on variability as a value opportunity?
- Have you explored the opportunities around predictive analytics for your operation?
- Is the accountability for data analytics in your organisation clear, or is it fragmented and misaligned?

3.8 FUTURE-PROOFING THE CHANGES

OBSERVATIONS

As we have discussed, taking the operational basics to a whole new level involves addressing a number of different areas: whole-system optimisation, linkages between planning/execution/improvement, integrated planning, disciplined execution, sustainable improvement, hard-wiring safety and risk management, and data analytics.

The changes required in each of these areas are common sense and essentially a refinement of what you are trying to do already. However, in combination, they undoubtedly represent a significant change to the overall way you may run the business at the moment.

The beauty of the changes is that they are all aimed at making it easier for people to do the right thing, in the right way, at the right time. If approached properly, it will build engagement and ownership, and remove friction and frustration. It is difficult to argue that you shouldn't be making these changes. They require some effort to put in place, but they are all things you should be doing anyway, and the rewards are considerable.

The holistic approach and attention to detail in the design is what makes the fully integrated operating model so different. It provides context; it clarifies the overall objective and everyone's individual roles; it provides systems and processes to help align everyone's activities; it improves transparency of information; it encourages collaborative behaviour; it improves delineation of levels of work; and it simplifies the working environment through

The difference between a CCR, an IOC and a fully integrated operating model

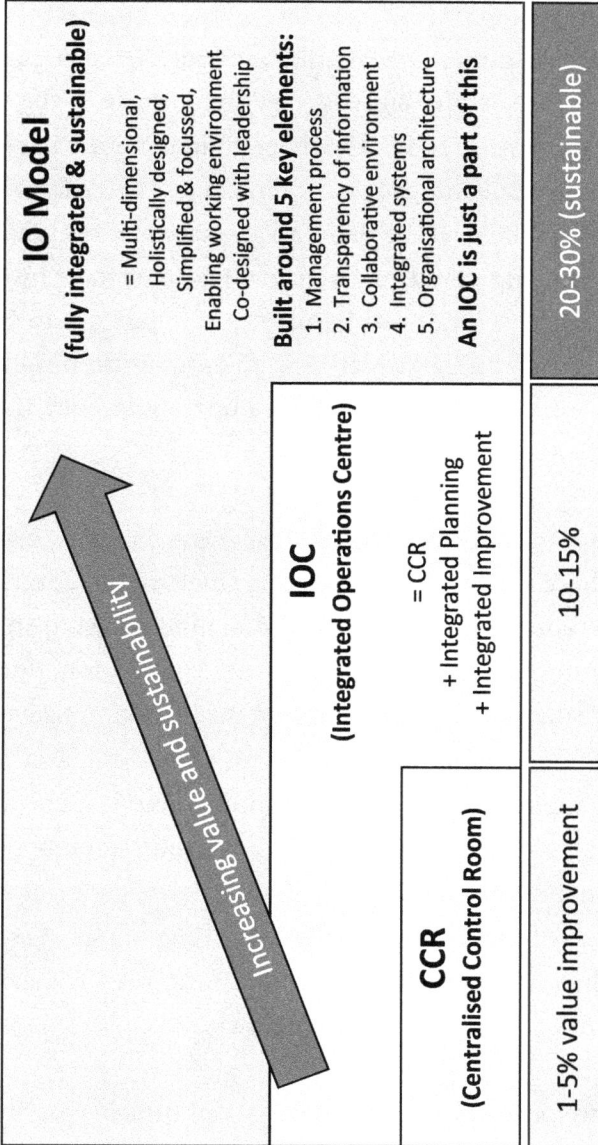

IO Model
(fully integrated & sustainable)

= Multi-dimensional,
Holistically designed,
Simplified & focussed,
Enabling working environment
Co-designed with leadership

Built around 5 key elements:
1. Management process
2. Transparency of information
3. Collaborative environment
4. Integrated systems
5. Organisational architecture

An IOC is just a part of this

IOC
(Integrated Operations Centre)

= CCR
+ Integrated Planning
+ Integrated Improvement

CCR
(Centralised Control Room)

Increasing value and sustainability

20-30% (sustainable)

10-15%

1-5% value improvement

Fig 3.8 Future-proofing the changes

decluttering of bureaucracy and introducing more intuitive, user-friendly designs.

These reasons in themselves should be enough to ensure sustainability, because sustainability is the core purpose of the whole integrated operating model. However, human nature is a strong force, and over time there is a risk of managers wanting to revert to the 'good old days' where they had greater autonomy, had less transparency, and could run their own silos at their own convenience. So, four other mechanisms are important to ensure these changes stand the test of time and are future-proofed: an Integrated Operations Centre (IOC), value transparency, training, and governance.

IOC: An IOC provides the 'home' for these changes, bringing together these different aspects into a single nerve-centre facility and enabling collaboration and dynamic decision-making. It houses integrated planning, integrated execution, dynamic decision-making, data analytics, and integrated improvement under a whole-system mandated view. It isn't essential to the success of the integrated operating model, but it is certainly a major enabler to functionality. And it represents a symbolic commitment to the transformational change being undertaken and the breaking of ties from the old way of doing things. As such, it is one of the most important future-proofing steps you can take in your journey to operational excellence.

Value Transparency: The most powerful future-proofing protection, however, will come from transparency of value-add. It is therefore essential that value-add from the integrated operating model (and specifically the IOC, if you have one) is understood, displayed, and reported, on an ongoing basis. Increasing performance trends, level of standardisation, variability, predictability, and other efficiency and value trends are all things that need to be continually held up to the light. If the integrated operating model is the heart of the operation, and the IOC is its nerve centre, it needs to be clear to all what value they bring.

Training: Training generally has weaknesses in many industries, but this certainly is the case in mining. Not technical training, or corporate training, which are usually done reasonably well. But practical leadership training, which teaches the art of leading and managing teams, and especially leading and managing an integrated mining system. Up-front and ongoing training is an important part of future-proofing the changes and needs to be given priority attention.

Governance: The other important future-proofing function is governance. Like all critical business functions, which are strongly governed, the same applies to the integrated operating model. Periodic audits should be carried out to ensure systems and processes are still happening in the way they should. This should include auditing leadership behaviour and workforce understanding. Auditing also serves as a means of understanding the relative strengths and weaknesses across the business, and where leading-practice transfer would be appropriate. It is also important to audit the quantification of ongoing value-add through the operating model.

Sadly, too often, businesses underestimate the importance of the holistic fully integrated operating model, and they think the construction of an IOC, a bit of centralisation, and partially integrated planning will change their world. It won't. It takes a lot more than that, as anyone who has gone down this path can attest to. However, the benefits are enormous for those who do it well.

PRINCIPLES

IOC: IOC design needs to be approached with caution. The physical facility is less important than the systems, processes, functionality, and mandates behind it. This is often misunderstood by companies who build a beautiful facility only to find that it hasn't added much value. If it isn't adding clear value, you haven't approached it right. If your goal is simply a building that looks good, with some centralisation of people, then lead with that. If your goal is to add value, then you need to start there. Where is the value currently won and lost in the operation over the short, medium, and long term? What obstacles stand in the way of realising that value? What do we need to do to fix those? And lastly, how can an IOC help enable and sustain this?

Value Transparency: Value-add is difficult to quantify, and this is sometimes used as an excuse not to do it. Don't accept this. Value quantification has a major role to play, and it should be a nonnegotiable requirement of the IO model and IOC management team to continuously capture value added through good decisions and improving performance trends. After all, it's their job to add value through their role. The precision of the value information is less important than the magnitude and recognition of value.

Training: Practical leadership training in the art of leading and managing in an integrated mining system is essential to realise the full value from an operation. It needs to form a part of the implementation-and-change programme. This needs to be carefully designed and delivered, as many of the principles will require good understanding and changes in behaviour.

Governance: What gets measured gets done. What gets periodically audited stays done. But it is crucial that this has a high profile. It should be recognised as fundamentally important to the future of the business. As such, it is important enough to be linked to the corporate governance level.

SOLUTION

The solution is to address all four of these areas.

Value, transparency, training, and governance are critically important.

An IOC has big advantages, and in some form is essential. However, this does not mean that the right answer for everyone is a large, fancy, custom-designed facility located in the nearest city. It needs to be fit-for-purpose and address a specific value proposition. In its simplest form, it can simply be a custom-designed room that houses planning, execution, and improvement staff with good transparency of information and some collaborative processes to drive additional value through integrated, whole-system thinking.

The application, design, and implementation of IOCs is a specialist area of expertise with a lot of factors to consider. Unfortunately, companies often underestimate the complexities of getting this right, and as a result, they don't leverage the full value from what they put in place. NextGenOpX can help you with this if you are thinking of going in this direction.

CHECKLIST

- How well-entrenched are your current business systems and processes, and would they stand the test of time if there were a high turnover within your leadership team?
- Do you have a good sense of the value that is won and lost in the interfaces between all the parts of the business?
- Have you assessed the value of a fit-for-purpose IOC strategy for your operations (you may start with just a simple colocation of planning, execution, and improvement roles)?
- Do you currently provide training to your leaders in how to lead and manage teams, and how to lead and manage an integrated mining system?
- How well are your business improvement systems and processes defined, managed, and governed?

CHAPTER 4

INTEGRATION PROVIDES THE MISSING LINK FOR TRANSFORMATIONAL CHANGE

4.1 Integration Is a Systemic Approach, Which Is What Underpins All Great Companies

4.2 Integration Provides a Structured Framework for a Systemic Operating Model

4.3 Integration Is an Objective, Measurable Approach

4.4 Integration Is an Untapped Area of Huge Latent Value

4.5 Integration Provides a Solid Platform for a Progressive Technology Strategy

4.1 INTEGRATION IS A SYSTEMIC APPROACH, WHICH IS WHAT UNDERPINS ALL GREAT COMPANIES

OBSERVATIONS

Jim Collins, in his famous book *Good to Great*, concluded that the key differentiator of truly great companies was their investment in a systemic approach.

The 'good' companies are continually changing and reinventing themselves. They have a 'flavour of the month' approach, and a haphazard approach to improvement, which means they are constantly redefining business priorities and chasing after new initiatives. They worry a lot about what others are doing and go after made-up time frames and targets. Their people feel the churn and are confused about exactly what is expected of them.

The 'great' companies understand what they are good at, and do that one thing exceptionally well, built around a clear operating model. They do it systematically and continually invest in improving the methodology behind it. They measure against themselves and are very clear on where they are going; they're always worried about continually improving, not on superficial time frames and targets.

The mining industry in general most certainly falls into the 'good' category—constantly changing organisation structures, and running major improvement initiatives, which in the long run barely move the dial. Very little that is systemic, and judging by the evidence of results, very little that is sustainable. Worse than that, periodically including ruthless cost-cutting initiatives, which often destroy the very fabric that had the potential to build

A systemic approach ... moving from good to great

"Investment in a systemic approach – the key differentiator of truly great companies"

(Reference: Jim Collins's book "Good to great", supported by comprehensive Harvard studies)

- ○ Essential for a sustainable solution
- ○ Stabilises the organisation
- ○ Enables continuous improvement
- ○ Knowledge is embedded

- ■ **Understand what they are good at**
- ■ **Do that one thing exceptionally well**
- ■ **Do it systematically**
- ■ **Continuously invest in improving the methodology**

- ■ Constantly changing and re-inventing
- ■ Flavour-of-the-month approach
- ■ Haphazard approach to improvement

Clear strategy
- ● Right value drivers
- ● Identify best practices

Performance inflection point
- ● Leverage synergies
- ● Realise option value
- ● Respond to external shifts

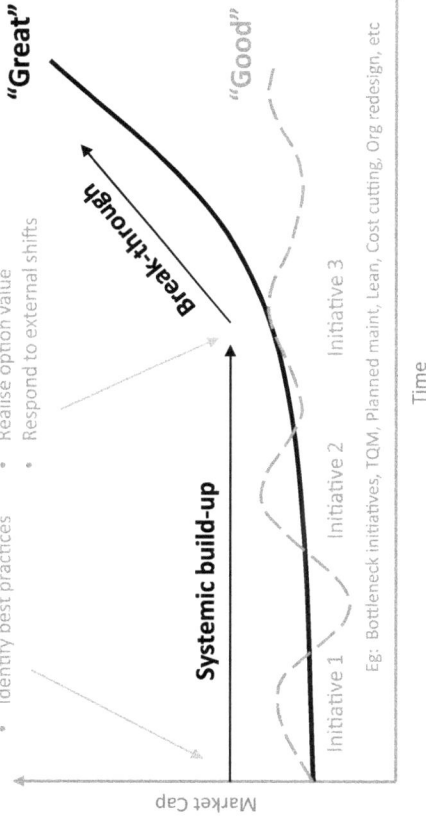

"Great"

Break-through

"Good"

Systemic build-up

Market Cap

Initiative 1   Initiative 2   Initiative 3

Eg: Bottleneck initiatives, TQM, Planned maint, Lean, Cost cutting, Org redesign, etc

Time

Fig 4.1 Integration is a systemic approach, which is what underpins all great companies

long term, systemic, sustainable value. The short-term value generated by these initiatives may make the project managers and associated consultants look good, but it can leave behind a trail of destruction in the business. This frustrates and exasperates employees, because it typically makes their lives more difficult.

The industry has tried in the past to tackle change through standards and systems, and now it's in danger of being seduced by the 'digital' silver bullet, which I have referred to previously. But all of these represent attempts to find a singular solution to what is actually a much greater multidimensional problem. A successful business is made up of a combination of good values, strategy, a management process, transparency of information, collaborative environment, integrated systems, organisational architecture, and culture. This is a complex mix. When these things are misaligned, you can't just fix that with a new system, or through introducing standards, or by digitising the misalignment.

What is needed is a methodology to align and simplify businesses and especially complex operations. A way of making the lives of employees easier. Making it easier for them to do the right thing, in the right way, at the right time. Liberating them to add more value rather than continually constraining them. The systemic methodology the industry is crying out for is integration. It is the underlying reason why the industry has failed to drive systemic, sustainable improvements in performance over the past two decades. It is the key to removing the obstacles that stand in the way of breakthroughs in performance, and accelerating/leveraging the full value of the next phase of the technological revolution.

There is one area where the industry has successfully achieved systemic, sustainable improvement, and ironically it is not the mining operations themselves; it is safety. Not the whole industry, but certainly a number of major mining companies can say they have truly achieved continuous, systemic, sustainable improvement in safety. Over a period of decades, they have demonstrated exactly that in their performance and their business culture.

Analysing this through the lens of integration, it is clear why. It is because the key principles of integration have been well-applied. A highly systemic, holistic approach. Clear strategy and expectations. Driven powerfully from the top. Strong, nonnegotiable standardisation with regard to management process, transparency of information, collaborative environment, integrated systems, and organisational architecture.

Inadvertently applied integration principles, but applied nonetheless. Perhaps it's because the methodologies were borrowed from other industries, which refined them empirically over time rather than understanding the principles behind them. So, we already have a perfect example of where integration has demonstrated its value.

Now, how about applying that same approach to the core business of mining? This is what the holistically designed integrated operating model has been designed to do. In one respect, it is a more complicated challenge than safety and has more moving parts, but on the other hand, it is a lot more tangible. Whatever the case, it is the right approach. The analogy of the journey on safety is a good example to use for understanding the magnitude of what is achievable if the industry puts its mind to it and approaches fully integrated operations in the right way.

PRINCIPLES

The key here is to turn the operation into a 'machine', which has worked out the best way of doing things and consistently applies that; continually learning, improving, and recognising that the whole is greater than the sum of the parts.

To build on the soccer analogy introduced in section 2.3, there are many examples of teams full of individual superstars that don't perform well. Similarly, there are teams full of just good players, who are unbeatable. The difference lies in commitment to 'team' over 'self'. It lies in collaboration. It lies in practice and repetition. It lies in systems and discipline. It lies in culture. You don't just pitch up to a big game; you develop a game plan, and you focus on executing that game plan in a way that every team member can rely on the other person doing their part. No heroes. Just a well-executed team result.

The mining industry can learn from this.

For every action there is a reaction. It is important to think things through when making changes. It doesn't make sense to do things that benefit you in the short term but create new problems for you later. In the mining industry, some examples of poorly-thought-through changes include retrenching experienced people, high-grading the ore body, reducing production stocks, reducing inventory levels of parts and consumables, reducing equipment flexibility, and removing/changing BI teams.

There are times, of course, when these changes are warranted, but too often they are done in a broad-brushed, blanket approach across an organisation without properly understanding the consequences. They very often don't achieve the desired goal in the long run. This is not systemic thinking.

There is a better way. Continually investing in systemic, sustainable changes, which make it easier for everyone to collectively perform better, is the right way to go. This will enable and motivate your teams to unleash the 20, 30, 40 percent of latent value that is currently stuck in the multitude of siloed interfaces across your organisation.

SOLUTION

The solution is to set up your business around a systemic approach.

Get clear on what your options, vision, and constraints are. Define the right operating philosophy for your operations. Design the right strategy, game plan, and roadmap to deliver this. Build this around a holistic-integrated operations framework. Align everyone and everything around delivering this strategy.

Make this the overriding focus of everyone in the business. Continually invest in improving and refining the model.

CHECKLIST

- Are you really clear on the purpose and competitive differentiator of your business?
- Does your entire workforce understand that?

- Is everything they do built around delivering that?
- How well have you defined 'The Way We Work' and the systems and processes that everyone should follow?
- Is everyone motivated to continually invest in improving these, systemically?
- Do you recognise and reward your people for being a hero and delivering great *personal* results, or for being a great *team* player who improves 'the system' for the long-term benefit of the organisation?

4.2 INTEGRATION PROVIDES A STRUCTURED FRAMEWORK FOR A SYSTEMIC OPERATING MODEL

OBSERVATIONS

Holistic design is the key to an integrated operating model. Notice the two important words: holistic and design.

Mining companies tackle a lot of 'stuff'. All the right things. But they tackle them in isolation of each other.

For example, they will tackle KPIs. There will be a beautifully executed programme to define the right KPIs and the right cascading of these through the business. They probably even include how the reporting processes need to change to align with these new KPIs. They may even ensure that these are embedded into the dashboards used throughout the workplace. A textbook piece of work.

But then they don't align this with the way the operators actually operate the mine, and the metrics they use in the front line. They don't adjust the way the plans and schedules are constructed and communicated. Or the way performance against these plans and schedules is communicated. Or communicate to everyone the importance of changing the terminology used by everyone in this regard. Or even fully explain the reasons for the change. Or seek feedback on the implications of the change for those in the front line. When different levels of management, right up to the corporate team, talk about the operation, they may still refer to the old terminology and metrics. The incentive schemes are not fully aligned with the new metrics. The data analytics processes are not aligned to the new metrics. The training systems are not changed to incorporate the new terminology and expectations. The documented standards and working procedures aren't changed, and the governance processes are not revised.

For the person responsible for doing the work on KPIs, they get a big tick and probably a good result on their bonus. But for the workforce who are left with a lot of misalignment with other messaging, systems, and processes, it just adds confusion and clutter.

The KPI example is just one of many that are happening continually. Whether it is new Lean or MOS (management operating system) processes, or reporting systems, or planning systems, or incentive schemes, or training, or organisational changes, the outcome is the same: an 80/20 approach to solving a specific problem where getting it 80 percent right takes 20 percent of the effort, and 'close enough is good enough'. This sounds logical, but what it fails to recognise is that each time this is done, it just adds a little bit more misalignment and confusion to the picture.

When it comes to value, the 80/20 approach makes sense. 80 percent of the value comes from 20 percent of the effort. But the problem is, we subconsciously apply the same principle to design. When it comes to design, the goal should be to get as close to 100 percent right as possible, even if it takes a lot more effort. Why? Because it affects everyone who uses the design forevermore. And because any little misalignments just add clutter and confusion to an already complex operating system.

Anyone who has ever used a cheap smartphone or a poorly designed app knows how frustrating that can be. We have zero tolerance when it comes to things like that. Yet, in the workplace,

Five key elements underpin NextGenOpX's approach to designing Integrated Operations

Holistic design of all of these elements is key, but there are a few game-changers

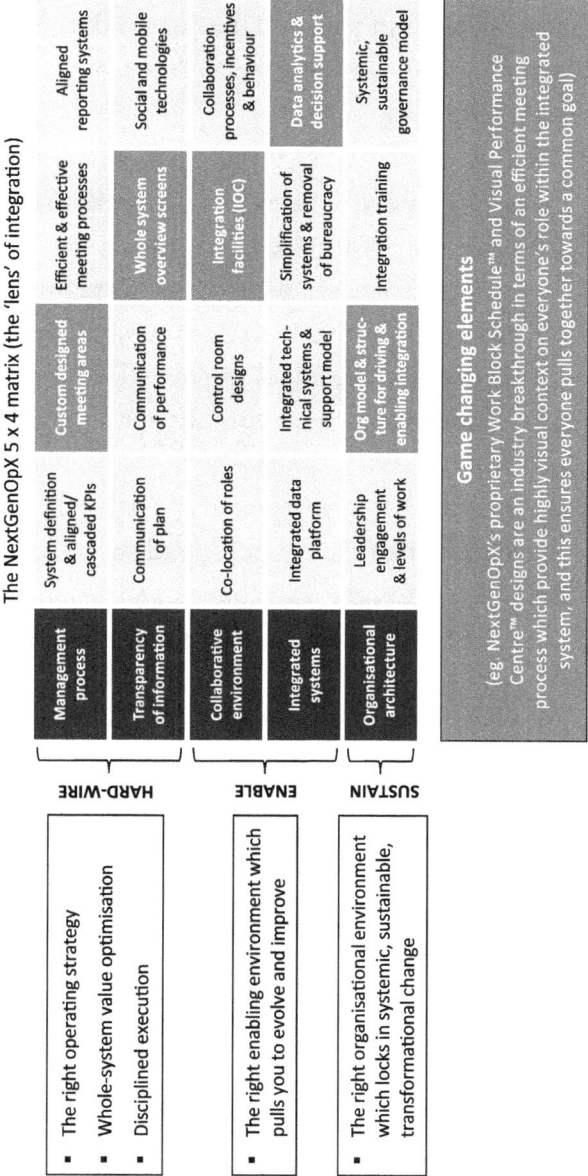

The NextGenOpX 5 x 4 matrix (the 'lens' of integration)

Management process	System definition & aligned/ cascaded KPIs	Custom designed meeting areas	Efficient & effective meeting processes	Aligned reporting systems
Transparency of information	Communication of plan	Communication of performance	Whole system overview screens	Social and mobile technologies
Collaborative environment	Co-location of roles	Control room designs	Integration facilities (IIOC)	Collaboration processes, incentives & behaviour
Integrated systems	Integrated data platform	Integrated tech-nical systems & support model	Simplification of systems & removal of bureaucracy	Data analytics & decision support
Organisational architecture	Leadership engagement & levels of work	Org model & struc-ture for driving & enabling integration	Integration training	Systemic, sustainable governance model

HARD-WIRE { Management process, Transparency of information }
ENABLE { Collaborative environment, Integrated systems }
SUSTAIN { Organisational architecture }

- The right operating strategy
- Whole-system value optimisation
- Disciplined execution

- The right enabling environment which pulls you to evolve and improve

- The right organisational environment which locks in systemic, sustainable, transformational change

Game changing elements

(eg. NextGenOpX's proprietary Work Block Schedule™ and Visual Performance Centre™ designs are an industry breakthrough in terms of an efficient meeting process which provide highly visual context on everyone's role within the integrated system, and this ensures everyone pulls together towards a common goal)

Fig 4.2 Integration provides a structured framework for a systemic operating model

we think nothing of implementing half-thought-through designs, or selecting the cheaper version of the new system we are implementing, which doesn't have all the bells and whistles that make it easier and more effective to use. We are happy to pass this 'clutter' down to the workforce.

As an industry, we need to become far more conscious of design, and far more dedicated to excellence in design, ensuring that the systems and processes we expect our teams to use are as user-friendly and intuitive to use as possible. They should not be designed to make the job of the owner of the system easier, but designed to make it easier for the user.

The framework for a fully integrated operating model is a holistic one. It describes all of the elements and subelements that need to be designed. The empirically developed 5×4 matrix shown in Figure 4.2 is built around five key elements: management process, transparency of information, collaborative environment, integrated systems, and organisational architecture. In combination, these form the 'lens' of integration (i.e. the way to look at your operation through the eyes of integration). Use it to design your systems and processes holistically around these five elements and twenty subelements, and design each with a view to simplifying the user interface.

A common response to this is, 'Where is integrated planning'? The answer to that is, 'If you want to do integrated planning well, you need to consider these five elements and twenty subelements'.

Another response is often. 'We are already doing all that'. Yes, but it has all been developed in isolation of each other. That makes all the difference in the world to the clarity and simplicity of the operating system as a whole.

PRINCIPLES

The above-mentioned integration framework with its five elements is crucial. It defines the holistic playing field, which needs to be considered in any fully integrated operating model.

Management process and transparency of information relate to establishing the right operating strategy, and hard-wiring the principle of whole-system optimisation, integrated planning, disciplined execution, and systemic improvement into the way the operation is run.

Collaborative environment and integrated systems are aimed at establishing the right enabling environment, which pulls you to evolve and improve.

Organisational architecture is focussed on establishing the right organisational environment, which locks in systemic, sustainable, transformational change.

Individually, they are just 'ho-hum' normal parts of running an operation. But when approached together and holistically designed with a focus on simplification, they are the key to transforming your business and releasing tens of percentage points of latent untapped value.

SOLUTION

Use the 5 × 4 matrix as a 'lens' of integration when thinking about how your operation, or parts of it, are currently set up and managed.

This allows you to see the business in a whole new light. It will suddenly become very obvious to you why the challenges you face are currently so hard to fix. In essence, you will likely see that the way your whole business is set up encourages fragmented, siloed, and misaligned behaviour.

The first step is to recognise this and to understand that it is fixable with a fully integrated operating model, when approached in the right way.

CHECKLIST

- How many of the things in the 5 × 4 matrix are you currently doing?
- How much effort is being put into the design (is the goal to be 80 percent right, or 100 percent right)?
- How well are each of the elements and subelements integrated with each other?
- Are they being tackled in isolation or being considered holistically?
- Is there an overarching game plan for designing and implementing all of this, and is it a staged journey?
- Who is managing this, given it is so critical to the success of the business?

4.3 INTEGRATION IS AN OBJECTIVE, MEASURABLE APPROACH

OBSERVATIONS

Benchmarking is an important process. However, too often the benchmark being chased is 'leading practice'. The problem with this is that leading practice may not actually be that good.

A commonly used scale is Poor—Average—Good—Leading Practice. This is subjective, and actually tells you very little, except how you are doing compared to your peers. That has some value, of course, but it doesn't speak to the full value potential you should be aspiring to.

An objective scale is much more powerful, one where the measures are absolute, where the scale doesn't change as the industry changes.

For productivity and performance data, the ultimate benchmark is the theoretical maximum. But for a subject area as amorphous as operating models, this becomes a much more difficult challenge. If it was possible to measure this in absolute terms, on a scale with clear points of definable delineation between levels of maturity, it would be extremely useful.

Fortunately, there is such a scale.

The internationally recognised Baldrige Excellence Framework includes an assessment scale that defines four distinct stages of business maturity.

1. **Reactive:** Many activities, few processes (reacting to problems; goals are poorly defined).
2. **Early systematic:** Beginning to conduct operations by processes with repeatability, evaluation, and improvement (early coordination among organisational units [e.g. business unit, department, section]; strategy and quantitative goals defined).
3. **Systematic and aligned:** Processes are repeatable and evaluated for improvement with learnings shared (coordination among organisational units; processes address key strategies and goals).
4. **Integrated:** Processes are streamlined and focussed with leading practice evaluation and transferred seamlessly (organisational units all supporting the system needs with collaborative behaviours; tight strategic alignment with key strategic pathways well-defined).

By applying these to the 'lens' of integration, i.e. each of the twenty subelements in the 5 × 4 matrix described in the previous section, NextGenOpX has created a highly effective Integration Maturity Assessment tool.

This is an absolute scale and the clear definitions remove a great deal of the subjectivity involved in assessing how good an operation's operating model actually is. It also clearly illustrates how much value potential is still left untapped in the way an operation is set up.

NextGenOpX uses the international Baldrige criteria for business excellence as a framework for a customised diagnostic tool for evaluating integration maturity

Reacting to problems

Strategic and Operational Goals

Operations are characterized by activities rather than by processes, and they are largely responsive to immediate needs or problems. Goals are poorly defined

Early systematic

Strategic and Operational Goals

The organization is beginning to carry out operations with repeatable processes with evaluation and improvement, and there is some early coordination among organizational units. Strategic and quantitative goals are being defined

Systematic and Aligned

Strategic and Operational Goals

Operations are characterized by repeatable processes that are regularly evaluated for improvement. Learnings are shared, and there is coordination among organizational units. Processes address key strategic and operational goals

Integrated

Strategic and Operational Goals

Operations are characterized by repeatable processes that are regularly evaluated for change and improvement in collaboration with other affected units. The organization seeks and achieves efficiencies across units through analysis, innovation, and the sharing of information and knowledge. Processes and measures track progress on key strategic and operational goals.

Visuals reproduced with the permission of the
National Institute of Standards & Technology (NIST)

Steps Towards Mature Processes: Assessment Considerations	Reactive	Early systematic	Systematic and Aligned	Integrated
	Many activities, few processes	Beginning to conduct operations by processes with repeatability, evaluation and improvement	Processes are repeatable and evaluated for improvement with learnings shared	Processes are streamlined and focused with leading practice evaluation and transferred seamlessly
	Reacting to problems	Early coordination among organisational units (eg dept, MRU, Section)	Coordination among organisational units	Organisational units all supporting the system needs with collaborative behaviours
	Goals are poorly defined	Strategy and quantitative goals defined	Processes address key strategies and goals	Tight strategic alignment with key strategic pathways well defined

Fig 4.3 Integration is an objective, measurable approach

"Fully integrated" is comprehensively described by NextGenOpX, including multiple examples and case studies with design detail

PRINCIPLES

The benchmarking principles behind the integration maturity scale are extremely solid.

Malcolm Baldrige was appointed to the position of US Secretary of Commerce by Ronald Reagan in 1981. In addition to his many achievements in the areas of international trade policy and trading practices, Baldrige played a leading role in building a renewed emphasis on quality and business excellence across multiple US industries.

In 1987, this culminated in the development of the Baldrige Business Excellence Framework. This falls under the auspices of the NIST (National Institute of Standards and Technology) in the US, and stands to this day as a leading repository of knowledge on business excellence. This includes a comprehensive management system for driving business performance improvement, and the associated Baldrige Award for those who achieve true business excellence.

By applying the detailed 'lens' of integration to this maturity scale, it creates a powerful new tool for businesses to benchmark their operating models.

This new integration benchmarking tool has been instrumental in exposing just how much opportunity there is for improvement in this space. It adds weight to the argument that there is typically 20–40 percent latent value currently lost to businesses, simply through the way their operations are set up. See Section 4.4 and Figure 4.4 for further detail.

SOLUTION

Carrying out an Integration Maturity Assessment on your business is an essential first step.

Why would you not want to get an objective benchmarking assessment of something as critical as your operating model? Especially when it is such an intangible thing and has probably received little scrutiny in the past, at least from a holistic-design perspective.

An assessment can be carried out in various forms: a simple online self-assessment (thirty minutes), a more rigorous facilitated assessment (two to three hours), or an independent on-site review (two to three days).

Whichever one you start with, it will get you thinking about your operations in a different way, and it will raise questions you won't have confronted before.

CHECKLIST

- Do you have a clear operating model (i.e. a structured way in which the business is set up)?
- Is this well-defined and communicated?
- Is it well-managed and governed?
- In your business strategy, have you considered to what extent your operations are enabled or hampered by the way they are set up?
- Can you say how well your operating model stacks up against the rest of the industry and your competitors?
- Have you identified areas that are holding you back and what value might be attached to addressing these?

4.4 INTEGRATION IS AN UNTAPPED AREA OF HUGE LATENT VALUE

OBSERVATIONS

As discussed in the previous section, the Integration Maturity Assessment tool assesses the five key elements of integration (and twenty subelements), against a spectrum of maturity ranging from Reactive, to Early Systematic, to Systematic & Aligned, to Integrated.

Applying this to mining companies across the industry presents a very interesting picture.

It shows that a large number of the operations are running in 'Reactive' mode (see Figure 4.4). The average spans Reactive and Early Systematic. The industry leaders are generally spanning Early Systematic and Systematic & Aligned. No one is sitting squarely in the Systematic & Aligned zone and, aside from a few subelements for a few companies, no one is in the Integrated zone.

This is quite a confronting picture when you consider the magnitude of the impact on the way operations are run. It goes a long way to explaining why the industry as a whole has struggled to drive sustainable productivity and performance improvement over the past two decades. In fact, it has generally seen a steady decline.

Interestingly enough, it does not come as a surprise to most operations people. When asked where their operation and the industry sit on this spectrum, they usually get it exactly right. This demonstrates how the frontline mining leaders of today are only too aware of the extent of the weaknesses in the way companies

The scale of the opportunity being missed by operations is illustrated by our diagnostic tool *

Benchmarking across the mining industry illustrates the enormous upside potential

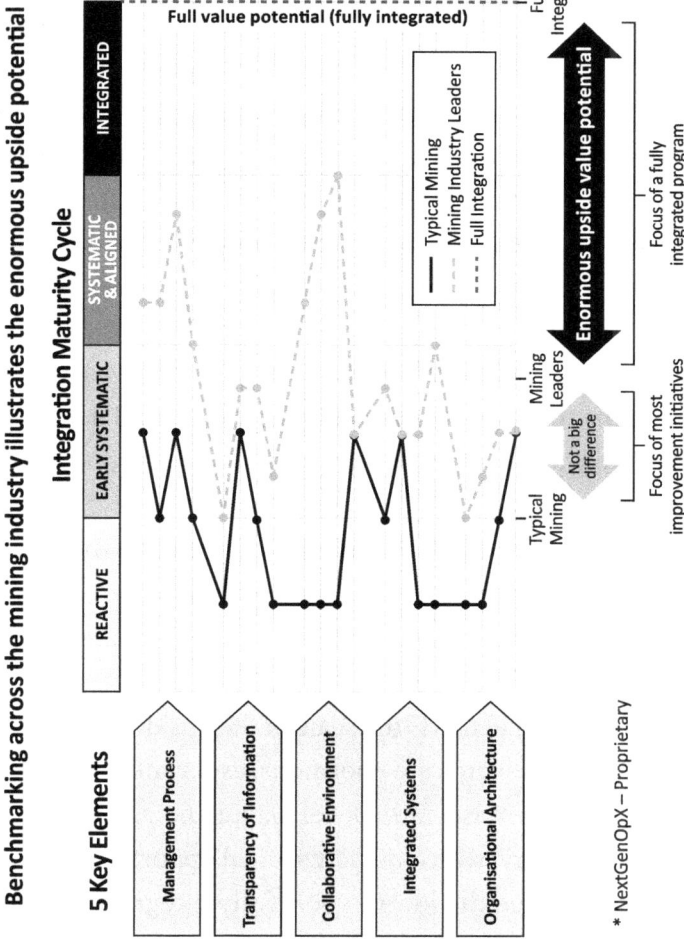

Fig 4.4 Integration is an untapped area of huge latent value

currently operate. It is a great frustration to them, working in such an environment every day, and they are very supportive of fixing these systemic issues. Operations people especially are generally very excited by the proposition of a 'better way'.

Senior executives are not always so clear on this. The information that gets filtered through to them is always trying to present a rosy picture. So, they can be forgiven for being under the impression that things are fantastic in their operations, and that everyone is crystal clear on what is expected of them, and all the teams are working to a unified drumbeat. The reality, of course, is that this is far from the truth, even in the best-run companies.

This is not to say the industry is terrible. It achieves incredible things in the toughest working environments, and it has a lot to be extremely proud of. It could just be a lot better.

When you are the best in the industry, or close to the best, it is easy to become complacent. But, as explained in the previous section, being the best doesn't necessarily mean you are very good.

The exciting thing in this is the scale of the upside opportunity. It is huge, and the rewards are enormous, especially for those who lead the way. There is a fairly level playing field at the moment across the industry, which presents a real opportunity for those who truly embrace the journey to a fully integrated operating model, to create step-change improvement for their business and leapfrog their competition.

For the sceptics out there who may believe the 'Integrated' zone is not realistically achievable in a mining environment, here is an interesting fact. Whenever we assess large operations using the Integration Maturity Assessment tool, there is always a huge range of results for the different parts of the business. Typically, there are some sections of the business that are completely 'Reactive' and rate a stone-cold zero on the scale; and in the same category (subelement) there are sections that have exceptional systems and processes in place and are well on the road to 'Integrated'.

This just goes to show that the expectations of 'Integrated' are not unachievable; they are actually being achieved today in parts of your operations. The problem is that they are not systemic, and no one is taking the trouble to identify them and replicate them across the business, and systematise that effort sustainably. What currently happens, in reality, is that when that inspirational leader moves to a new role, they take that methodology with them, and the new leader who replaces them drops it like a hot potato. This is because they don't understand it, have never done it before, and are often working in a culture where strong leadership means you call the shots from the top and drive your results with a 'big stick'.

There is no rocket science in this. There are many actions we should be embracing enthusiastically, not debating whether they are worth doing: aligning everyone and everything around delivery of the business strategy; simplifying/clarifying/aligning/integrating the working environment to make that easier; and investing the time and effort to do that systemically and sustainably.

The playing field is currently open, and the opportunity is there for the taking.

PRINCIPLES

The Integration Maturity Assessment process is not intended to be an end unto itself. It is simply a tool that helps you see where your business or operation currently sits with regard to your operating model. It also shows how different parts of your business compare, and how you compare with other businesses.

When approached thoroughly, it can provide you with considerable insights into your operations and help you understand the interconnectivity between the five elements: management process, transparency of information, collaborative environment, integrated systems, and organisational architecture.

But perhaps one of the most important benefits is that it helps you to identify those pockets of leading practice that currently exist within your business and to recognise the leaders who are likely to be the ones you need to be investing in more.

Occasionally, you will come across a crusty old battle-hardened, 'old-style' leader...the 'big stick, loud voice' type who says, 'I don't need all this integration stuff; I know how to get the job done; I can get results'. And they are probably right, in the short term. But what they don't appreciate is that when they leave, everything will fall in a heap, because everything relies on *them* to get things done properly. When confronted with this fact, the reply is usually, 'Well, that's not my problem; I've done *my* job'.

And there, in a nutshell, lies the problem. Or two problems, to be precise. First, they don't care about the 'whole system'; they just care about their little piece. And second, they don't consider systemic improvement and sustainability to be a part of their job.

That is why this subject is so important to address. How can you possibly hope to build, run, and grow a great business when this is the mindset? And it has become a very prevalent mindset in the absence of a better way.

SOLUTION

The key is to recognise that you need this different approach. Then to move on it. And then to follow it through to the end.

It is not an easy process to make things easier for your teams to perform, but for any leader, this should be a primary goal: not to micromanage them, but to enable them.

The greatest friction in a business lies in the interfaces between different people who are pulling in different directions. The answer lies in cleaning up the operating model, simplifying the working environment, focussing on whole-system optimisation, integrated planning, disciplined execution, fit-for-purpose standardisation, and building a culture where an improvement is not viewed as an improvement unless it is a systemic, sustainable improvement; and strengthening the organisational accountabilities around delivering this outcome.

CHECKLIST

- Have you ever objectively assessed how good your operating model is?
- Does your business/operation focus on systemic, sustainable improvement, or is it all about short-term results?

- Do you see your organisation repeating mistakes over the years and doing a lot of reinventing of the wheel?
- How different would your business/operation be if you could fix these issues, and how big would the prize be?
- If you had a clear strategy and roadmap for achieving this, would you go after it?
- Where would the resistance come from? Are you sure?

4.5 INTEGRATION PROVIDES A SOLID PLATFORM FOR A PROGRESSIVE TECHNOLOGY STRATEGY

OBSERVATIONS

Most companies nowadays have a technology strategy. Technology is so obviously a part of the strategic equation for any business, it is hard to ignore it.

Defining the technologies that will have a great impact on your business in the future is a more difficult challenge. Should your business play an instrumental role in the development of those technologies, or should you be a 'fast follower'? And if you choose to be a 'fast follower', is that an actual strategic position, or is it just a euphemism for 'I don't really know what I'm going to do'? Which may not necessarily be a bad position to take, when you consider how uncertain the future pace and direction of new technologies actually is.

Once a new breakthrough technology has been developed and proven, the time it takes for this to become widely used is relatively short. So, is there really that big of a competitive advantage in being at the forefront of new developments?

These are all interesting questions, and there are probably no magic answers. It is likely to be very dependent on the style of the company, the necessity for the breakthrough, the nature of the technology, and whether it is likely to come from the private sector in due course anyway.

However, what is clear in the mining industry is that the 'technology' focus has for a long time largely been centred on mining equipment technologies per se. That is, things related to automation, like autonomous trucks and drills, continuous rock-cutting machines, decision-support software, condition-monitoring sensors, advanced exploration techniques, etc.

The initial emphasis is on making the technology work. It is only when the technology is proven, and attempts are made to implement it into an existing or new operation, that the next hurdle is revealed: how to actually make it work within an operating environment that is used to a completely different paradigm.

Experience with the implementation of autonomous trucks was a prime example of this (already raised in section 1.4). Once the technology was proven, it would be a simple matter of rolling it out to different operations, right? Wrong. The actual evidence showed that there was an enormous difference between the experience of different mines, even within the same company. Some showed major improvements in productivity (30 percent), others showed virtually none at all. It all had to do with the way it was approached by management, and the way it was adopted into the operation. Basically, change management and the level of ownership engendered in the workforce were what made the difference.

Another good example is mining software. There are some amazingly powerful systems being produced nowadays, but somehow they never quite live up to expectations. This is not a function of their capability. It is a function of the way they are applied. When you speak to the leaders of these software companies, they absolutely concur with this conclusion, and they express a level of frustration at not being able to overcome it. In their words, 'We

You can't just jump to Technology / Digitisation / Automation – it needs staging

Any technology strategy should target 3 horizons (all influencing each other and happening concurrently)

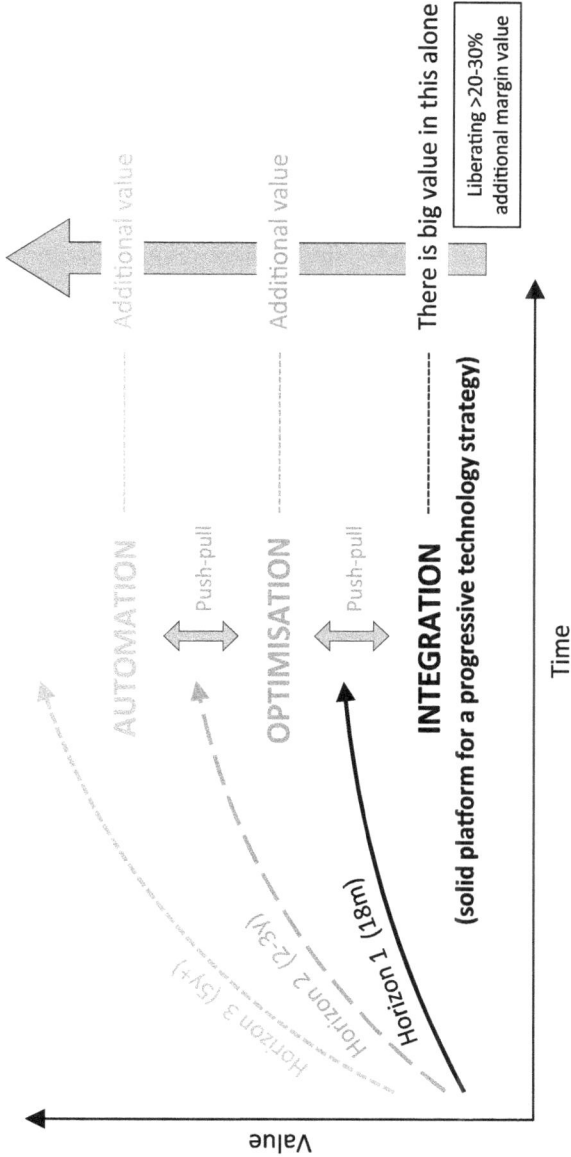

Fig 4.5 Integration provides a solid platform for a progressive technology strategy

are good at design and implementation, but we are not good at *adoption*'.

In other words, the working environment into which these technologies are inserted is the deciding factor on value-add. It is the organisational readiness that ultimately counts. The people interface.

Therefore, it makes sense to focus more on the enabling platform for technology. If it isn't completely clear exactly where technology itself is heading, then it makes even more sense to create the operating platform that will inspire the development of the technologies, and that will enable the rapid adoption of them when they arrive. It makes no sense to just sit back and wait for the technologies to arrive, only to suddenly find out at that point that you are completely unprepared to adopt them rapidly into your organisation.

A fully integrated operating model, which is clear, optimised, aligned, disciplined, and simple to follow, is an essential starting point. It will pay for itself quickly, regardless of other technologies. And you can start working on it now. If done properly, it will not only prepare the ground for new technologies and digital, but it will also create a hunger for it within your organisation.

Don't wait until the technology arrives before you move. It may be too late then, and you will find yourself scrambling to adjust while others are leapfrogging the curve.

PRINCIPLES

There is a lot of talk about 'digitisation'. Not surprisingly, there is plenty of confusion about it, because it is an extremely broad concept. The concept is fine, because it means moving away from clumsy and disconnected analogue information, to an age where all information is available instantaneously, everywhere, and can even be used for algorithmic control, and ultimately control by artificial intelligence.

That end-point position sounds great, but how do you get there? The problem is that some consultants are persuading companies that this will fix all the dysfunctionalities they are currently experiencing in their operations. The message conveyed is: 'All you need to do is digitise everything and then you can bypass a lot of the people-involvement that so often gets in the way of efficiency. Once it is digitised, you can then automate it, and your problems go away.'

As mentioned previously, this is missing the fact that digitising (and automating) a poor operating system just gives you a highly repeatable poor operating system. At some stage, you actually have to sit down and improve the design of what is done. Consultants often give the business leaders a very reassuring message on this: 'We won't dictate to you what to do in this regard. You tell us the right processes and we will digitise/automate them'. This is another way of saying, 'We don't know how to do this; we hope you do'. And of course, the operations people haven't been able to figure this out themselves to date, otherwise they would have done so, and the industry productivity and performance statistics would be a lot better than they are.

So that puts us back to square one. You need to deal with the operating model first, and improve the way the basics are carried out. This then beautifully paves the way for digitisation. Not superficial digitisation, where various digital projects are patched together and forced into a highly analogue organisation and culture. But true, deep, transformational digitisation.

SOLUTION

Get to grips with your operating basics now. It is a no-regret decision.

Don't put your faith in any proposal that suggests digitising everything up front will magically fix things for you. In other words, don't digitise your rubbish! On the contrary, getting the basics right will strengthen and accelerate your ability to take on the new technology when it comes. At the very least, do both in parallel.

A well-designed, holistically integrated operating model will help you deal with today's issues, as well as create the platform for a progressive-technology strategy.

This needs to be approached as a transformation journey, which is owned and led by the leadership team. It involves designing and implementing a carefully staged and integrated roadmap, which requires running three horizons of work concurrently (the initial basics of integration; the enablers that bridge the present and the future; and the future technology/digital strategy itself). It is not simply a matter of running a bunch of separate project initiatives.

CHECKLIST

- Do you have a clear technology strategy?
- Is the burning platform behind it well-understood by the workforce?
- Do all your leaders (especially the operational leaders) believe in it, embrace it, own it, and champion it (or do they just passively support and endorse it as someone else's agenda)?
- Does it include the horizon one and two enablers, namely the integration and optimisation phases, as precursors to the horizon three technological future?
- Are these three horizons integrated into a single coordinated roadmap?
- Is it being managed purely by the technology teams, or do the operational leaders have a firm hand on the rudder?
- Is this complex, integrated programme being managed and driven by a powerful, high-profile programme management office (PMO)?

CHAPTER 5 ///

A FIVE-STEP GUIDE FOR INTEGRATING YOUR BUSINESS

///

5.1 Approach It as a Journey, Not a Big Bang, and Closely Engage Your Leadership

5.2 Develop a Staged Roadmap, Which Delivers Quick Wins but Also Longer-Term Goals

5.3 Customise This to Align with Your Business Strategy and Existing Initiatives

5.4 Build the Capability of Your Leaders to Lead This and Upskill Your Workforce

5.5 Follow a Comprehensive Integrated Operations Framework for Implementation

5.1 APPROACH IT AS A JOURNEY, NOT A BIG BANG, AND CLOSELY ENGAGE YOUR LEADERSHIP

OBSERVATIONS

Delivering systemic, sustainable, cultural change in a business is essentially a transformation process, and it needs to be managed as a journey, not a 'big bang' implementation.

The key is to bring along the hearts and minds of the workforce. When new methodologies or technologies are involved, it is easy to lose sight of the fact that the real challenge lies in adoption and leveraging the full value from the change. Therefore, close engagement of the leadership team is essential, and this is best done through a codesign process.

It is undoubtedly a change management challenge. However, no beautifully crafted change management process can make up for a poor design, whether it is the design of the product or the design of the implementation approach. Dilbert said it best:

There are many problems with trying to implement a transformational change in one big step. It usually involves a long study, a big design team, potential overdesign, less engagement with

A practical pathway to a smarter, simpler and more productive operating platform

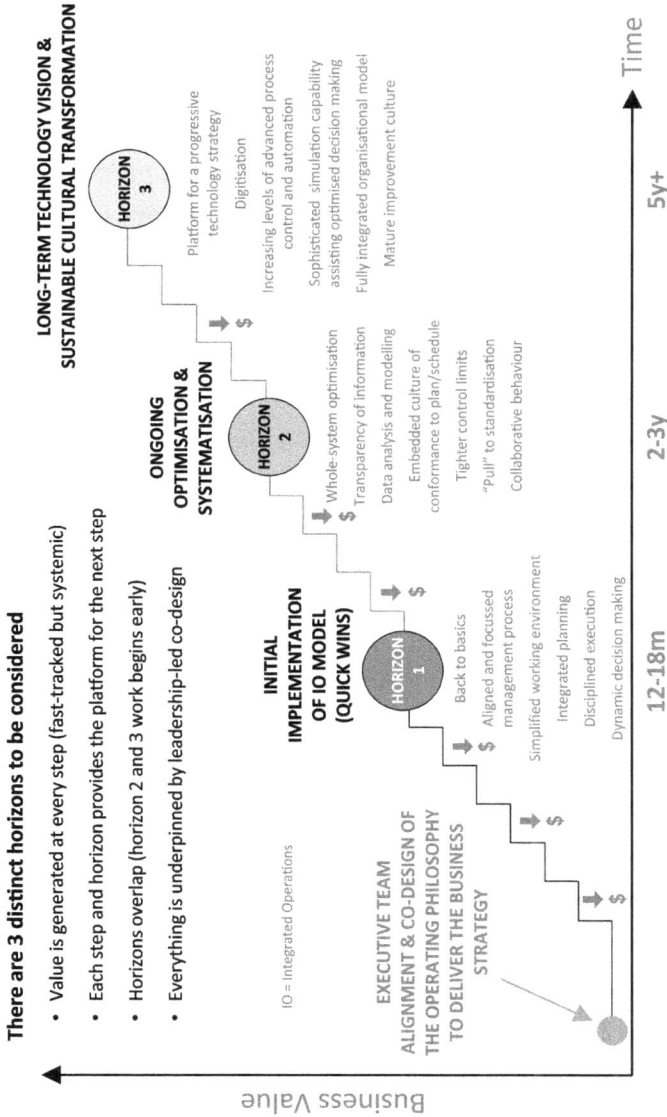

There are 3 distinct horizons to be considered

- Value is generated at every step (fast-tracked but systemic)
- Each step and horizon provides the platform for the next step
- Horizons overlap (horizon 2 and 3 work begins early)
- Everything is underpinned by leadership-led co-design

IO = Integrated Operations

EXECUTIVE TEAM ALIGNMENT & CO-DESIGN OF THE OPERATING PHILOSOPHY TO DELIVER THE BUSINESS STRATEGY

INITIAL IMPLEMENTATION OF IO MODEL (QUICK WINS)

HORIZON 1

Back to basics
Aligned and focussed management process
Simplified working environment
Integrated planning
Disciplined execution
Dynamic decision making

ONGOING OPTIMISATION & SYSTEMATISATION

HORIZON 2

Whole-system optimisation
Transparency of information
Data analysis and modeling
Embedded culture of conformance to plan/schedule
Tighter control limits
"Pull" to standardisation
Collaborative behaviour

LONG-TERM TECHNOLOGY VISION & SUSTAINABLE CULTURAL TRANSFORMATION

HORIZON 3

Platform for a progressive technology strategy
Digitisation
Increasing levels of advanced process control and automation
Sophisticated simulation capability assisting optimised decision making
Fully integrated organisational model
Mature improvement culture

Business Value

Time

12-18m   2-3y   5y+

Fig 5.1 Approach it as a journey, not a big bang, and closely engage your leadership

the operations team, higher risk, a much bigger people challenge, and the vision is often a moving target.

Fast progress helps drive the change and creates an inspirational 'pull' for the business. So, a much better approach is to stage the implementation through pilot trials, codesigned and delivered by identified passionate 'champions' within the operational teams. This builds engagement and ownership and leads to learning by doing, which demonstrates the vision to the workforce rather than just describing it. This is a much lower-risk approach, which leads to fast results and quickly builds momentum and confidence.

The leadership team must codesign the vision and roadmap to ensure full ownership, understanding, and commitment to the journey. It is not enough to have a study team work out the plan, and then present it to the leadership team for endorsement. That simply provides approval to proceed with 'your project'. It doesn't achieve the deep understanding and ownership of 'our transformational programme, that I will lead from the front', which is essential for successful transformational change.

The vision and roadmap should be built around three horizons:

1. **Integration**—short-term priorities (getting the basics right, establishing a stable/repeatable system, and getting everyone and everything pulling in the same direction, around the right business strategy and operating philosophy).
2. **Optimisation**—medium-term enablers (starting to optimise the stable system and creating the right enabling environment that pulls everyone towards the technological/digital future).
3. **'Automation' (Digitisation)**—longer-term technology vision/

strategy (the technology and digitisation projects that are required to realise the long-term sophisticated business strategy, including a sustainable cultural transformation).

PRINCIPLES

The focus of each of the three horizons is as follows:

1. **Initial implementation of the IO model (quick wins):**
   - Back to basics (getting the operating fundamentals right)
   - Whole-system operating philosophy
   - Aligned and focussed management process
   - Simplified working environment
   - Integrated planning
   - Disciplined execution
   - Dynamic decision-making
2. **Ongoing optimisation and systematisation:**
   - Whole-system optimisation
   - Transparency of information
   - Data analysis and modelling
   - Embedded culture of conformance to plan/schedule
   - Tighter control limits
   - 'Pull' to standardisation
   - Collaborative behaviour
   - Fit-for-purpose Integrated Operations Centre (IOC) strategy
3. **Long-term technology vision and sustainable cultural transformation:**
   - Platform for a progressive technology strategy
   - Increasing levels of advanced process control and automation

- Sophisticated simulation capability, assisting optimised decision-making
- Fully integrated organisational model
- Mature systemic improvement culture

SOLUTION

The important thing with this is to start with codesigning a clear vision and implementation roadmap with your leadership team. Awareness of leading practice and emerging technologies in the mining and other industries is important. The key is to maintain the right balance between stretching the aspirations and grounding the vision in realistic practicality. So, there is a need for understanding the future technologies, combining that with a deep operational empathy, and being able to facilitate a diverse leadership team to the right conclusion.

The integrated operating model is a solid approach for horizon one.

Leveraging the platform created in horizon one and enabling the pull to horizon three is the focus of horizon two. A fit-for-purpose Integrated Operations Centre (IOC) strategy is often the catalyst and game changer for locking in this phase. Building a strong data analytics capability is also a feature.

With the solid platform created by the horizon one and two work, the new technologies in horizon three are able to much more seamlessly lock into place, having been developed, trialled, and proven in parallel with the horizon one and two work.

CHECKLIST

- What is your own experience with transformational programmes (successes and failures)?
- Do you understand the reason why some succeeded and some failed?
- Did the organisation learn from the mistakes or continue to repeat them over time?
- Do you have a clearly defined technology and digital strategy?
- Does your leadership team own and drive this as if it was theirs, or is it viewed as the VP of technology's programme, which they 'support'?
- Does it stand alone as a technology strategy, or is it structured as a staged transformation journey?
- Have you started with a comprehensive leadership engagement step up front, to shape the vision, design the roadmap, and define the right implementation approach, in which everyone has a part to play?

5.2 DEVELOP A STAGED ROADMAP, WHICH DELIVERS QUICK WINS BUT ALSO LONGER-TERM GOALS

OBSERVATIONS

Too often, the focus of a technology plan is all on the technology itself, and the journey to get there receives too little attention.

In reality, identifying the technologies and even describing the end-point vision is not actually that hard. Designing the right staged journey, however, is a more complex exercise, yet extremely important to get right. A well-structured and carefully designed roadmap significantly accelerates results and reduces risks.

Staging the journey ensures early gains will be built on a solid foundation and not slip back.

As discussed in the previous section, there are three horizons of work (integration, optimisation, and automation), which essentially relate to the short-, medium-, and long-term time frames, respectively. These should be executed in parallel, not in a series, but must build progressively, so that the one horizon provides the solid platform for the next horizon.

The time frames are dependent on the nature and size of the business/operation. However, as a general rule, horizon one can be thought of as the twelve-to-eighteen-month time frame; horizon two as the two-to-three-year time frame; and horizon three as the five-years-plus time frame.

A structured roadmap for an industry leading integrated operating system connected to the digital pathway will create the game changing opportunity

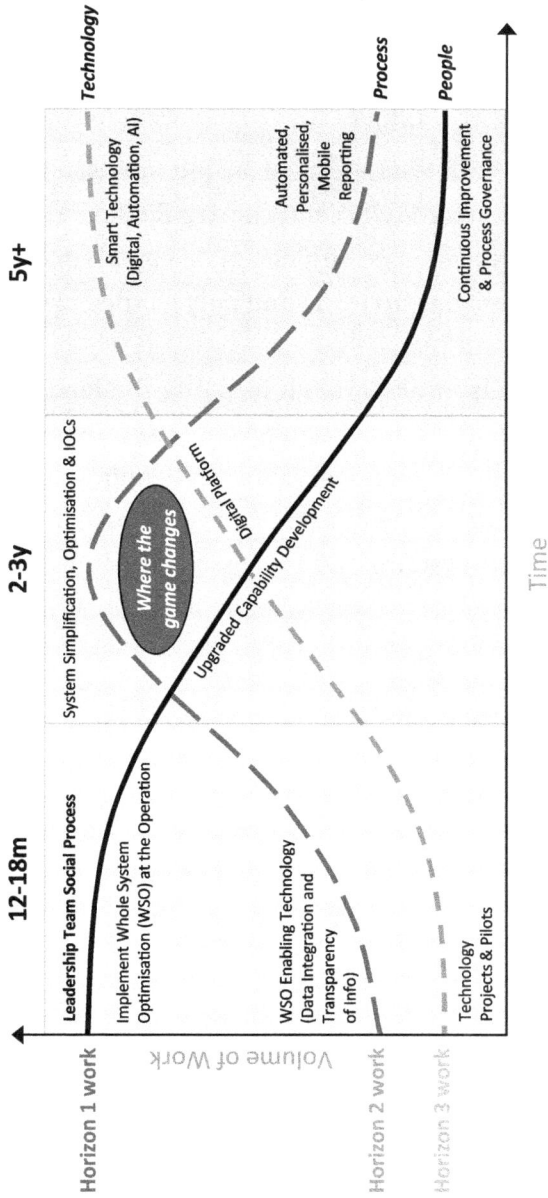

Volume of Work (y-axis)

Time (x-axis)

Horizon 1 work
Horizon 2 work
Horizon 3 work

12-18m

Leadership Team Social Process

Implement Whole System Optimisation (WSO) at the Operation

WSO Enabling Technology (Data Integration and Transparency of Info)

Technology Projects & Pilots

2-3y

System Simplification, Optimisation & IOCs

Where the game changes

Digital Platform

Upgraded Capability Development

5y+

Technology

Smart Technology (Digital, Automation, AI)

Process

Automated, Personalised, Mobile Reporting

People

Continuous Improvement & Process Governance

Fig 5.2 Develop a staged roadmap, which delivers quick wins but also longer-term goals

Figure 5.2 shows the way these horizons and time frames should progress together. During the twelve-to-eighteen-month time frame, the greatest volume of work is on the horizon one work, with some horizon two work, and a little horizon three. In the two-to-three-year time frame, the greatest emphasis is on the horizon two work, with the horizon one work starting to decline, and the horizon three work starting to build. In the five-year-plus time frame, the greatest emphasis is on the horizon three work, with some horizon two work, and only a little horizon one work.

Horizon one is initially about building the leadership team's social process and implementing whole-system optimisation as the focus of the operations. Then really upgrading the capability development of the leaders and workforce, to leverage the value from the systems and processes put in place. And finally, moving to an emphasis on systemic improvement and process governance.

Horizon two is initially about defining and designing the technology and functionality, which will enable whole-system optimisation, including things like data integration and transparency of information. Then simplifying and optimising the systems and processes throughout the business, and establishing the right fit-for-purpose IOC strategy. And finally, moving towards automated and personalised mobile reporting.

Horizon three is initially about defining the technology projects and starting to develop, test, and run pilot trials. Then really starting to digitise the business in a holistic and structured way that enhances the horizon two work and enables the horizon three work (not random disconnected projects). And finally, moving towards more advanced smart technologies, like full digitisation, automation, algorithmic control, and AI.

The horizon two work in the two-to-three-year time frame is when the game really changes and the business truly starts to shift to a new paradigm of work. Usually at this stage, the operational stability, coupled with increased analytics capability, and rapid dynamic decision-making creates options for changing the way certain parts of the operation are set up and run. That's when the true shift in value delivery takes place. There are quick wins to be had right from the early months, but these culminate in a step-change synergistic shift at around the two-to-three-year time frame.

PRINCIPLES

There is an important distinction between horizons and time frames, and it is important not to confuse the two.

Horizons relate to the staging period when the work has greatest value, emphasis, and impact. Essentially, the horizon one work is the work that needs to be done up front to set up horizon two, which in turn sets up horizon three. Although horizon one work is largely completed in the twelve-to-eighteen-month time frame, it doesn't stop at that point; it just naturally trails off as the next horizon builds up.

Time frames are the timing of when the work is carried out. All three horizons of work involve work in the twelve-to-eighteen month, two-to-three year, and five-year-plus time frames. It is just the emphasis and volume that changes in each time frame. This is best illustrated in Figure 5.2.

The key is to design an interwoven set of carefully designed projects and milestones, which come together at the right time, to deliver clear outcomes, with each building on the one before.

When done well, this becomes the master plan for transforming your business, and it can be a powerful mechanism for unifying your leadership team and driving your business forward with clarity and confidence.

SOLUTION

Start with the leadership-led codesign process. Spend some time working out 'Where are we now?', defining what works well, what doesn't, and what the specific challenges are that you want to address.

Then move onto 'Where do we want to be?', defining your aspirational vision in specific terms. Don't constrain yourself during this process or find reasons why you can't do things, but do ground your thinking in the realms of the possible.

Then move onto 'How do we get there?', thinking through the aspects of the business/operation that need to change, specifically what needs to be done and what the key milestones are. Define the milestones you need to achieve over three horizons: twelve-to-eighteen months, two-to-three years, and five-years-plus.

Then develop the programmes of work behind each of these, and make sure the implementation plans and deliverables gel together well with each other. Confirm accountabilities and resourcing requirements for delivering the individual projects.

Most importantly, define who will own and manage the overall programme. The sponsor should be the senior leader of the business unit as a whole (could be CEO, MD, or GM level, depending on the size of the business unit), and this leader should play an active role in leading, driving, and designing the game plan.

Establish a programme management office. Develop a master schedule for all of the work, and manage the programme tightly, in a way that drives ongoing collaboration, accountability for key milestones, and unity of purpose and outcome.

You will likely need some external help in setting this up, at least for facilitating the vision and three-horizon strategy, and for developing the right fit-for-purpose integrated operating model for your business. However, you should not fully abdicate the design and delivery of this to a consulting company. You need to be the clear owners and drivers of the programme. Choose your delivery partner with that in mind. Operational empathy is an important quality to look for.

CHECKLIST

- Is your business vision and strategy an integrated one, or is it simply a set of loosely coordinated individual initiatives?
- Do you have a clearly assigned accountability for managing the collection of projects, and a process that drives alignment, collaboration, and unity?
- Do you find that some projects are well-delivered but don't actually move the dial meaningfully?
- Is there good alignment in expectations between the projects

delivered by the support functions and the operational team users who will be on the receiving end of them?

- Do you find that you often take two steps forward and one step back each time a new initiative is delivered?
- Would you describe your team as 'One team, with one goal'?

5.3 CUSTOMISE THIS TO ALIGN WITH YOUR BUSINESS STRATEGY AND EXISTING INITIATIVES

OBSERVATIONS

By design, a fully integrated operating model will be completely aligned with the overall business strategy. The very purpose of the integrated operations (IO) programme is to align everyone and everything around delivery of the business strategy. Yes, the process of delivering the IO programme may help you to sharpen that strategy, or maybe it won't; but it will certainly help you to achieve it.

There are always a number of local and corporate initiatives already ongoing in any business at a particular moment in time. Assuming they are adding value and remain part of the strategy, there is absolutely no reason why these should conflict with an IO programme of work. On the contrary, if well-designed, an IO programme should enhance any existing initiatives, not hinder them.

The IO programme process will help to identify and address any existing gaps and overlaps between activities across the organisation, including any challenges with existing initiatives.

There is sometimes a comment from leaders that, 'We have too much on the go already; this sounds good, but we need to finish what we are currently doing first; we'll come back to it when we are ready'. This is usually a sign that the organisation would benefit from better integration right now.

Not always, of course, and there can certainly be issues of mixed messages if too many things are done at once. But in truth, even in such cases, it still makes sense to start with the basics of IO. These don't have to be branded into a major new IO programme that could conflict with the existing messaging of another major transformation initiative. But undoubtedly, the basics of IO will be addressing existing gaps that will hamper you in the future, and these should not be delayed unnecessarily.

Putting IO off 'until you are ready', is, in reality, just a way of saying you'll wait until you are forced to do it in the future. It is not proactive. It is not wise. And the overload of issues you are currently experiencing (both literal and psychological) is usually a sure sign that you have a lot of clutter, confusion, gaps, and overlaps in the business, and you would really benefit from the radical simplification that is possible through a well-designed IO programme.

It is worth remembering that the unique focus of integration is not so much on the parts of a business but on the interfaces between the parts, as well as the way the whole system is optimised. There is nothing in the principles of clarification, collaboration, simplification, integration, and alignment that should interfere with existing initiatives. And, to the extent that you need to establish certain no-go areas within the IO programme, with good logical reason, there is no problem in principle with this.

Regardless of what you are currently busy with, it is worth at least exploring the opportunities and benefits of starting to integrate your operations more effectively.

A well-designed integrated operating model will align with and enhance existing initiatives

- **An integration programme won't conflict with existing initiatives:**

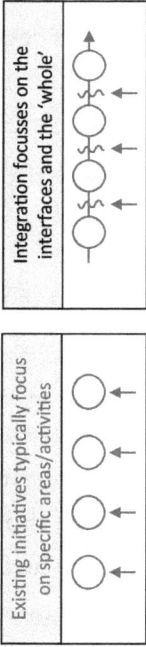

Existing initiatives typically focus on specific areas/activities	Integration focusses on the interfaces and the 'whole'
○ ○ ○ ○	○○○○○○

- **Integration is about aligning everyone and everything around delivery of your business strategy:**
  - It doesn't dictate that strategy per se
  - The process may help you to sharpen your strategy, or maybe it won't, but it will certainly help you achieve it

- **There is no reason to put it off 'until there is less going on':**
  - There will always be lots going on
  - Fear of mixed messaging is usually a sign that there is already clutter and confusion
  - Whether you use it as the primary vehicle for change, or just adopt some of the principles, it will <u>always</u> be better than not considering integration at all
  - The principles are simple and common sense, and will bring clarity and structure to complicated situations

- **Integration is not something to fear, it is a practical approach which has been proven to work in transforming the culture and performance of operations**
  - You need it most when there is a lot going on

Fig 5.3 Customise it to align with your business strategy and existing initiatives

PRINCIPLES

There are always a multitude of ideas and projects, in various stages of evolution, going on across an operation. One of the common dangers found in most operations is that there are far too many to manage effectively. There are a few ways to deal with this.

One is to simply cut down to the top ten (or whatever is the appropriate number), and ban work on anything else. This stifles initiative and can mean good projects get squashed and value is lost.

Another is to prioritise. This has sound logic and is certainly to be encouraged, but doesn't necessarily rationalise the number. I have seen examples where rigorous prioritisation exercises lead to greater clarity on the real issues, and businesses end up with more projects than they had started with!

Another is to introduce an exhaustive project management process, one that captures every project and forces it through a rigorous process until the point of completion. This is usually accompanied by a 'big stick' process to force leaders to make this a priority. This typically doesn't end well in the long run. It just creates a massive list of projects (sometimes in the thousands!) and the process starts to take on a life of its own. The frontline leaders get forced to deliver on these, and they end up spending much more time in the office on admin than out in the field. This leads to them taking their eye off the ball with regard to operations and safety, with negative results and a great deal of frustration.

Many of the ideas and projects have huge overlaps and implications for each other, so it is oversimplistic to think that you can just address a long list of things, and eventually all the problems will go away. In reality, solving one problem very often creates another elsewhere.

The answer to this is to really get behind the root cause of the problems you are trying to solve. That is, the systemic issues that underpin many symptoms (each of which currently end up getting defined as a separate project). The lens of integration, and the principles of IO, are an ideal vehicle to help you do this. And they will lead you down a path where you get the basics right and stop having to chase your tail putting out fires caused by underlying issues.

SOLUTION

A good starting point is to carry out an assessment of all the projects and initiatives, including new systems and processes, which are currently being managed and implemented in your business at the same time. Then, think about the gaps, overlaps, and potential confusion that may be resulting.

This should have a particular focus on the impact to the frontline operational workforce, who are the ones actually making the money for the business. It is important that the systems and processes being used, and the initiatives being implemented, are made as easy as possible for these frontline teams to use and implement. Unfortunately, this is often not the case, and it is common for systems and processes to be designed more from the perspective of the manager of the system than for the user. People in support-function roles should never lose sight of the fact that their job is to support and enable those frontline teams, not to hinder them.

Whatever the case, any IO programme should clearly identify all current major initiatives and programmes. It should understand their purpose, intent, and desired outcomes. It should ensure that in the process of delivering the other IO outcomes, it should enhance and enable the existing initiatives and programmes as far as possible.

CHECKLIST

- How many programmes and projects do you currently have running?
- How many of these have the aim of 'helping' the operational teams?
- How many actually end up helping operations, as opposed to distracting them?
- Have you tried to evaluate how much overlap and confusion exists between all of the projects and initiatives currently on the go (including those from the recent past)?
- Have you surveyed your leadership team and workforce to find out what they think about all the projects and initiatives (i.e. which they like and which they are confused about)?
- For the really big and important projects and initiatives, is anyone currently doing anything concrete about identifying and addressing the issues, gaps, overlaps, and confusion that exists?

5.4 BUILD THE CAPABILITY OF YOUR LEADERS TO LEAD THIS AND UPSKILL YOUR WORKFORCE

OBSERVATIONS

Most of the challenges we have been discussing throughout this book relate to leadership in some way: either recognising the need for transformational change, or the practical realities about how to address current operational weaknesses, or how to deal with a workforce within a socio-geopolitical environment that is rapidly changing.

At the heart of all these changes lies a people challenge, which quickly translates to a leadership issue, as leaders are the ones who tend to end up as the 'meat in the sandwich', so to speak. What is crystal clear is that the challenges for leaders at all levels in the mining industry (and other industries) are only going to get much harder over the next decade or so. And this is happening at a time when leadership 'capability' in general is struggling.

The bottom line is that there is a huge and growing need to better prepare leaders for the next generation. The journey ahead represents a paradigm shift for leaders and the workforce, and leaders can't be expected to navigate this without good quality training. There is now a greater need than ever to improve the quality of leadership and systematise the integration that used to happen naturally.

Current leadership training is generally pretty good on company values, corporate responsibility, and the importance of the environment, community, and diversity. However, what is missing is practical leadership training that accelerates the experience curve and equips leaders with real-world useful tips and tricks on the art of leadership: things they can take back to the workplace and apply immediately. Wise insights from successful mining leaders who have learnt from experience what to do and what *not* to do. We still see a lot of leadership theory, and 'death by PowerPoint', presented by support staff, in information-packed two-day courses, which leaders find increasingly difficult to fit into their already busy schedules.

Businesses need a much more efficient and effective way of training, that directly targets the current weaknesses being seen across the industry. These are: the basics of how to lead and manage teams and how to lead and manage an integrated mining system. This combination will equip them with the leadership skills required to inspire the next-generation workforce, give them the tools to drive the step-change to true operational excellence, and prepare their teams for the rapidly approaching next phase of technological revolution.

This is a major gap at the moment. Younger leaders need help and guidance to accelerate their path to becoming the leaders they want to be. Older leaders sometimes need to understand that their style, which used to work, may not do so in the current and future environments. Without a doubt, the training of leaders needs to improve in quality, content, delivery method, and experience for the attendee.

Building the capability of your leaders is an essential element of the transformational shift that needs to take place.

Leadership training is an essential element of a comprehensive integration strategy

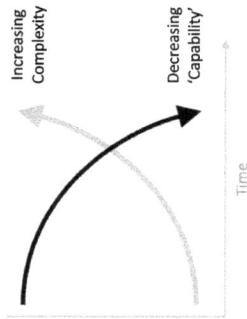

Increasing Complexity

Decreasing 'Capability'

Time

The challenge for the industry:

- How to address significant operational challenges
- At the same time as preparing for the next phase of technological revolution
- And to do so with a workforce, and within a rapidly changing socio-geopolitical environment, which is rapidly changing

The current weaknesses which need to be addressed:

- Many of the new generation of leaders struggle with the **basic art of leadership**
- The **growing demand for greater integration**, is a paradigm shift for leaders
- Leaders are facing **transformational change**, which they need to know how to drive

Leadership training required:

CULTURE

Leading and managing teams:

Your role, how to manage a high performing team, and how to build collaboration:

- Lead, motivate, coach and mentor
- Accountability and collaborative culture
- Meetings and communication

PREDICTABILITY

Mining as an integrated system:

How to successfully run a high performance mining operation as an integrated system:

- Whole system optimisation
- Clarity, focus, simplification
- Systemic, sustainable improvement

VALUE

Delivering step-change improvement:

How to implement transformational change in your company/mine/area:

- Context and understanding
- Engagement and ownership
- Performance culture

Fig 5.4 Build the capability of your leaders to lead this, and upskill your workforce

PRINCIPLES

In the mining industry, people are trained in how to do their jobs. But as they transition from someone responsible for doing their work to someone responsible for their teams doing the work, the expectations of their role change significantly. And yet, businesses are not properly training their people on how to make this change.

There are many issues with existing training. It isn't usually focussed on coaching leaders in the art of leadership. It usually requires at least two days of dedicated time, which is hard for operational leaders to fit into their schedules. It typically involves a lot of theory, presented in the form of multitudes of slides, which is known not to be an effective medium for learning. It is also often delivered by support staff who have not experienced the challenges of leadership themselves.

Leadership training needs a new approach. It requires real-world, practical training on how to lead and manage. Digital training, which can be completed in flexitime—a combination of knowledge-sharing, self-reflection, and interactive discussion within a peer group who are doing the course at the same time. Finally, the content should be thought-provoking and stimulating, facilitated by subject-matter experts and rich on practical insights and tips, which can immediately be applied back in the workplace to help students become better leaders.

It needs to be a scalable and sustainable training model, delivered on a top-quality digital training platform, in a user-friendly way, aimed at delivering a behaviour-changing experience. It needs to be learning that is fun and interactive, transparent, and insightful, engaging, and outcome-focussed. Otherwise it won't stick.

Addressing the leadership capability gap is a must-do. It is already an issue, and it will only get worse.

The starting point is to recognise that you have a problem, and to understand the needs and constraints for your particular business. Perhaps an open and honest survey on the quality of your existing leadership training would be an easy first step. This should test the waters on the issues raised above.

You will need to decide whether you develop this training from scratch yourself, or partner with someone who has something ready-made (which can be refined to suit your particular needs, including terminology and emphasis). It is worth thinking this through carefully, as many large companies have tried to do this themselves in the past, without great success. What tends to happen is that it morphs into something different to the original needs, and gradually fizzles out when changes in leadership occur. There is some serious merit in recognising that training is not generally a core competence of mining companies (especially digital training), and making the decision to outsource this.

In the absence of anything suitable available elsewhere, Next-GenOpX has developed custom-designed courses that tick the boxes on all of the points covered in the Principles section above. These target both leading and managing teams, and leading and managing an integrated mining system. They combine the operational experience of NextGenOpX in the content with the training delivery experience of a world-class digital training provider, and represent a very different and better experience for the participants.

Once you are confident that the quality of content and delivery is right (an important assumption to check!), then it means you can tick the training box and know that it will be carried out systemically and sustainably long into the future. It is much better to have a well-governed and quality-controlled system than have to revisit it with a new team every several years and have to start all over again.

Training of the quality required for mining companies moving forward is a specialist skill and warrants being treated as one of your highest-priority business systems.

CHECKLIST

- Are you happy with your current leadership training?
- Do you know exactly what content is taught and the delivery method?
- How well is it regarded by your leaders?
- What percentage of your existing leaders (at all levels) have received the training?
- How often is refresher training carried out?
- On average, how long does it take before new leaders receive the training?
- What is the true, full cost per attendee of the courses you deliver?
- When was the last review carried out on your leadership training content, delivery method, and attendee feedback?
- Would an online digital course be of interest to you (if at the same time it addressed current weaknesses)?

5.5 FOLLOW A COMPREHENSIVE INTEGRATED OPERATIONS FRAMEWORK FOR IMPLEMENTATION

OBSERVATIONS

To summarise the earlier chapters and sections so far: the industry has a problem with significant structural weaknesses that need to be addressed. The operational basics need to be taken to a whole new level. Integration provides the missing link for transformational change. You need to approach this as a well-structured transformational journey, which is strongly led by your leadership team.

There are multiple dimensions to this journey, which all have to be carried out in an integrated way, some in sequence, some in parallel. The principle of holistic design is crucial.

In short, it sounds like a daunting task, and it is certainly not a straightforward undertaking. But all of the elements for success are actually common sense and are mostly things that some enlightened individuals in your business are already doing parts of anyway (albeit in different and nonstandardised ways). So, in reality, what is missing is the knowledge that it can be done successfully, the courage to take on the change, and the catalyst to do so.

A strong case has already been made for the burning platform for change, so the catalyst and necessity is there. The integrated operating model provides the structured approach and framework for doing this, and is built on sound, proven principles and methodologies, applied with deep experience and operational empathy. So, there is no doubt it can be done.

That leaves the courage to do so. This clearly needs to be done. It will be hard work, but your business has likely tackled and succeeded with much greater challenges. If done properly, your workforce and leadership will love the change. It won't be 'just another initiative' that is forced upon them. It will be a transformation they will help to shape, design, and implement. And its very purpose is to make it easier for them and their teams to do their jobs. It will change their lives for the better.

It will need to be set up, driven, and managed in a way that holds true to the above values. It will need to be resourced and led in an appropriately powerful manner with a sustained commitment to follow it through to completion.

However, the value proposition of doing this is so large, and the risks of being unprepared for the coming next phase of the digital revolution are so great, that it would be fair to say that you really can't afford not to.

It needs to be done. So, you either do it proactively and benefit from the advantages of being an early mover or you do it reactively and find yourself scrambling to catch up with the industry leaders later on. What is certain is that generations of leaders after you in your organisation will thank you for the role you and your team played in making their lives easier.

In terms of making the decision to move on this, it is worth reflecting on the words of a famous saying: 'If not you, then who? If not now, then when?'

The NextGenOpX framework for designing and implementing integrated operations

The pathway to sustainable operational excellence

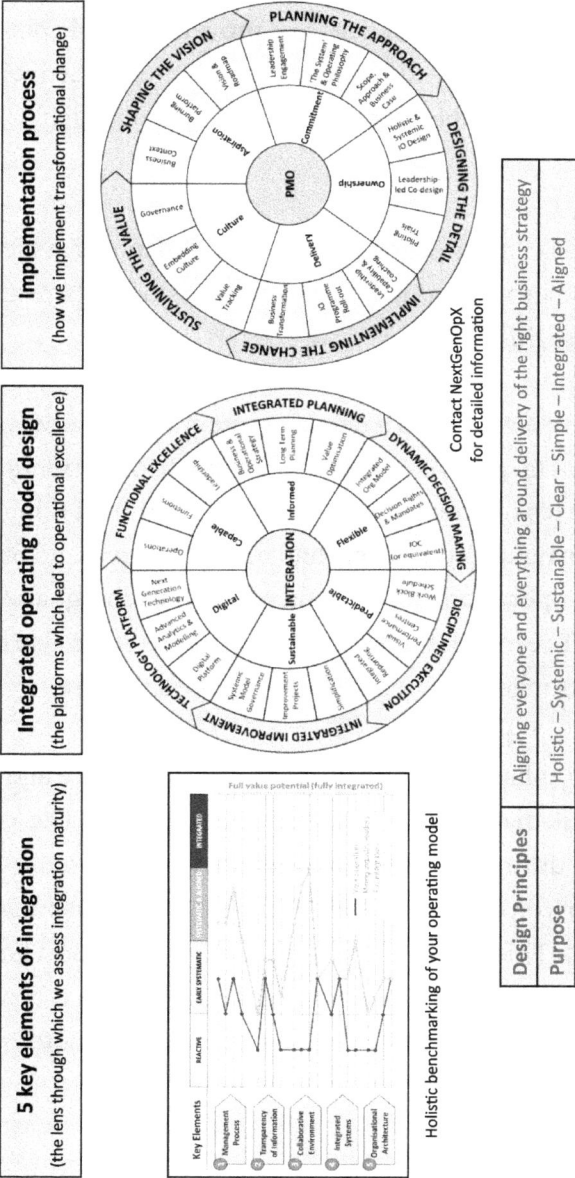

5 key elements of integration
(the lens through which we assess integration maturity)

Integrated operating model design
(the platforms which lead to operational excellence)

Implementation process
(how we implement transformational change)

Implementation process wheel (PMO)

SHAPING THE VISION · PLANNING THE APPROACH · DESIGNING THE DETAIL · IMPLEMENTING THE CHANGE · SUSTAINING THE VALUE

Aspiration · Commitment · Ownership · Delivery · Culture

Vision & Roadmap · Leadership Engagement · The Systemic & Operating Philosophy · Scope Approach & Business Case · Holistic & Systemic IO Design · Leadership led Co design · Piloting Trials · Coaching & Capability · IO Programme Rollout · Business Transformation · Value Tracking · Embedding Culture · Governance · Business Context · Building Platform

Integrated operating model design wheel (INTEGRATION)

INTEGRATED PLANNING · DYNAMIC DECISION MAKING · DISCIPLINED EXECUTION · INTEGRATED IMPROVEMENT · TECHNOLOGY PLATFORM · FUNCTIONAL EXCELLENCE

Capable · Informed · Flexible · Predictable · Sustainable · Digital

Business & Operational Strategy · Long Term Planning · Value Optimisation · Integrated Org Model · Decision Rights & Mandates · IOC (or equivalent) · Work Block Schedule · Planning · Improvement Projects · Systemic Model Governance · Digital Platform · Advanced Analytics & Modelling · Next Generation Technology · Operations · Function · Capability

Contact NextGenOpX for detailed information

Full value potential (fully integrated)

Key Elements:
Management Process · Transparency of Information · Collaborative Environment · Integrated Systems · Organisational Architecture

REACTIVE · EARLY SYSTEMIC · INTEGRATED

Holistic benchmarking of your operating model

Design Principles	Aligning everyone and everything around delivery of the right business strategy
Purpose	Holistic – Systemic – Sustainable – Clear – Simple – Integrated – Aligned

We have comprehensive detail behind each of these (ie Lens of Integration, IO model Design, Implementation process) backed by standards, design criteria, case studies and leading-practice examples

Fig 5.5 Follow a comprehensive integrated operations framework for implementation

PRINCIPLES

There is no rocket science in this. It is all built on solid logic, leading-practice experience, and proven methods put together into a practical, staged approach, custom-designed with operational empathy to be fit-for-purpose for your particular business needs.

You may ask if the integrated operating (IO) model is such a game changer, then why has it not been raised before? There are several answers to this.

First, there was always a need for integration, but when leadership teams (at all levels) used to have much longer tenure in roles, and were less distracted by other nonoperational things, it used to happen naturally. Leaders had more time to spend on optimising their operations through face time, coaching, and walking the talk. The longer tenure in management teams also helped to ensure stronger bonding and knowledge transfer.

Secondly, it has required a unique journey to arrive at the simple framework that lies behind the current IO model. The ideas were spawned during the design and delivery of the ground-breaking Rio Tinto Iron Ore Operations Centre in Perth, Australia, in 2008 and 2009. This courageous undertaking was a world first on anything like this scale in the mining industry, and it broke down many previously entrenched paradigms about remote operations and central-integrated planning mandates.

However, it took a few years and dozens of different applications across multiple continents, cultures, and commodities to understand that the true missing link was not an Integrated Operations

Centre (IOC); it was integration per se. Then it took a few more years for the generic principles of integration to fall into place, out of the diversity of applications tackled.

During the process of solidifying the IO model, it became increasingly obvious that it was worthless unless you could deal with the challenges of shifting the mindset of a senior leadership team, and carry them through the transformation journey. So, incorporating this into the methodology was another key step forward. Building all of this with deep, practical operational experience was also essential to success.

In summary, the keys to cracking the code on this were a combination of the rigour of the IO model and the leadership-led codesigned transformation process, designed and supported by experienced operational leaders with deep operational empathy.

Last, the devil is in the detail of the design. It has taken many applications to flesh out the design detail behind each of the five elements and twenty subelements of the IO model, and to ultimately test this all holistically in fully integrated operating model applications through major transformation programmes.

So, the 'how' of a fully integrated operating model has now been established and proven, including the key generic principles and design detail behind it.

SOLUTION

A structured, comprehensive approach is essential, using a holistically designed integrated operating model.

Some full-time internal resourcing will be required, along with a well-managed PMO (programme management office).

Some external help with integrated operations expertise is highly advisable, but the transformation process should not be 'farmed out'.

It is important that the codesign and implementation planning is approached with operational empathy and strong mining experience.

CHECKLIST

- Have you defined, assessed, and critiqued your existing operating model?
- Are you clear on the value proposition for change?
- Do you understand the risks and challenges you will face and need to overcome?
- Does your transformation programme include these elements?
  - Leadership-led codesign of the vision and roadmap over three well-aligned horizons (and time frames) of work
  - A holistic design, based on the five key elements of integration (management process, transparency of information, collaborative environment, integrated systems, organisational architecture)
  - Strong ownership by the leadership team as a whole and clearly defined individual accountabilities
  - A well-resourced PMO with an efficient master schedule process for maintaining close collaboration around the single end goal

- A fully aligned leadership development training pro-gramme
- Do you have the right experienced people in place to help you make this a resounding success?

CONCLUSION

///

The next phase of the technological/digital revolution is coming fast, whether you like it or not, and it will not be fully in your control. We live in an ambiguous and fast-changing environment. Leaders need to position their businesses based on the approaching trends. The fact is that businesses and operations (certainly in mining) are generally poorly prepared for this change. They are too focussed on short-term results and are not putting enough effort into establishing the solid operating platform necessary to successfully underpin future technologies and social changes. Without this foundation, businesses leave themselves vulnerable to ongoing declining performance and ultimately competitive disruption.

Advances in new technologies are currently running way ahead of our ability to adopt them, and this gap is growing. There is lots of focus (and hype) around the technology/digital space, but very little attention is given to the operating fundamentals, which are declining. The missing link to bridging this gap is integration. The mining industry has got away for many years without a structured approach to this, because integration used to happen naturally twenty years ago when operations were simpler and management teams were more stable. However, the combination of increasing complexity of operations and the decreasing tenure of leadership teams means that integration now needs to be systematised.

Many of the current weaknesses in the mining industry are directly related to integration: siloed behaviour, lack of whole-system thinking, prioritising the short-term urgent over the long-term important, standardisation, discipline, and lack of continuous sustainable improvement. Until now, the industry has written these off as 'just the nature of mining', partly because there has not been a viable alternative. But it *isn't* 'just the nature of mining', and there *is* a better way. The answer lies in building a fully integrated operating model.

The essence of such an operating model is quite simple. It is to get really clear on the right business strategy and whole-system operating philosophy, and then align everyone and everything in the business around delivering that strategy. The key is to holistically design the model around five key areas: management process, transparency of information, collaborative environment, integrated systems, and organisational architecture. There needs to be a particular focus on the interfaces between roles and functions, and the way everything comes together to optimise the overall value of the business as a whole. The end result is greater clarity for everyone in the business: better context; improved focus on the right things; increased alignment; simpler and more user-friendly systems and processes; greater transparency and accountability; and less friction, misalignment, and fragmentation of effort. The question needs to be asked: why would you *not* want to do this?

It requires some effort to set up properly, but the benefits are enormous, as there is typically around 20–40 percent latent value stuck in the interfaces across an operational business, as well as in the way the 'whole system' is optimised. So, the potential for transformational value-add for any such business cannot be ignored, not only in addressing today's problems but also positioning for the technological/digital future. This is not rocket science; it is all practical, doable, and proven in practice. However, it does need to be approached in a structured way with operational empathy, strong engagement of leadership, and simplification in mind.

This book has not only described how to go about building such an integrated operating model, but it has also laid out the case for why it is so important, and especially so at this moment in time. It tells a compelling and comprehensive story in five chapters with examples, principles, solutions, and checklists for easy reference.

I have explained why the industry has a serious problem and is in a vulnerable position. I have discussed the structural weaknesses that need to be addressed. I have provided detail on the operational basics that need to be taken to a whole new level. I have explained why and how integration provides the missing link for transformational change. And I have concluded with a five-step guide for integrating your business.

Hopefully this has given you cause to think differently about the way your business and operations are set up, and has motivated you to tackle something that has been ignored for too long.

If you are hesitant to take on this challenge, remember that your people on the front line have been craving this type of smarter operating system, and they can see the opportunities being lost every shift, every day. They have been stuck in a vicious circle where they are always reacting to yesterday's problems, explaining why they didn't make target and hoping they can just make it through the week without too many problems. Tackling integration in the way described in this book will bring light to the root causes of their problems and will address the systemic issues that have been holding businesses back from sustainable improvement. Integration is a recipe for making the lives of your leaders and workforce better and reducing frustration and waste.

With strong leadership intent and a well-designed integrated system, the adoption and acceptance by your workforce will not be the issue, provided you bring your team along on the journey. Allow your leaders and workforce to codesign with you, and create an environment where creativity, diversity of thought, and innovation are embraced, with a realisation that you won't always get it exactly right the first time. If you approach it in the right way, you will see unheard-of levels of engagement and ownership in the outcomes, and you will experience a deep cultural shift in your organisation.

Don't find reasons to put this off to a later stage when you 'have more time' and have 'finished dealing with some other urgent priorities'. The reason you don't have time and are flat out with other priorities is in large part because your business is unstable and your people are in 'churn' mode. Integration will help you deal with this. It is a common-sense approach, which will improve your performance and make life easier for your leaders and workforce. That, in itself, is a reason not to delay.

Integration can be tackled in multiple ways and designed to align with whatever existing strategies and initiatives you may already have ongoing. What is clear is that it will target additional untapped latent value, which these other initiatives are not fully addressing. Whether you start with a simple Integration Maturity Assessment, a small pilot, or you target specific aspects of the integrated operating model, or you tackle it as a comprehensive transformation, you should make a start. You will not regret doing so.

Any operation that aspires to true operational excellence should be following this approach. And any operation that intends to move boldly towards a digital future shouldn't think of doing so without doing this first.

SUMMARY

///

INTRODUCTION

- The mining industry is facing a period of major change, which it is not ready for.
- There is huge value at stake.
- This book presents a well-thought-through 'better way' of setting up operations built around a key weakness, which is integration.
- It is based on more than a decade of extensive industry-leading experience in this space.

CHAPTER 1: THE INDUSTRY HAS A SERIOUS PROBLEM AND IS IN A VULNERABLE POSITION

1.1 Productivity within the Industry Is Declining

- Productivity is declining.
- What we're doing isn't working.

1.2 Operations Are Not Getting the Basics Right

- Many of the fundamentals of operational excellence are missing.
- Things are getting worse.

1.3 Technology Is Growing Exponentially Fast

- Advances in technology are increasing rapidly on multiple fronts.
- There is a lot of hype in this space.

1.4 There Is a Big Gap Emerging That Is Not Being Addressed

- Advances in technology are running way ahead of our ability to adopt them.
- Technology is not a silver-bullet solution for poor operating fundamentals.
- We are missing something that provides a bridge between the today and tomorrow.

1.5 Businesses That Don't Adapt Face the Real Risk of Disruption

- The value opportunity in this gap is so great that someone will step in to seize it.
- New globally disruptive technologies will be the likely catalyst for this.
- Those who don't prepare well could be left standing by these technological changes.

CHAPTER 2: STRUCTURAL WEAKNESSES NEED TO BE ADDRESSED

2.1 Complexity Is Increasing and Capability Is Decreasing

- Mining operations are becoming substantially more complicated to run.
- 'Capability' of leadership teams is declining due to the impacts of shorter tenure and experience.
- In combination, these two factors represent a profound change, yet we continue to run operations in the same old way.

2.2 There Is More Business 'Clutter' and Less Clarity of Purpose

- Leaders and employees are not playing to a single, clear game plan within businesses, which leads to misalignment and confusion.
- Little attention is given to improving bureaucratic, user-unfriendly systems and processes that detract from value-adding work.
- Clarifying, aligning, integrating, and simplifying the business would make it easier for teams to do their jobs in the right way.

2.3 Behaviour Is More Siloed than Collaborative

- Siloed behaviour is the norm in the mining industry and this causes a lot of lost value.
- Our businesses are currently set up in a way that drives that behaviour.
- There is a clear need for a more collaborative and integrated approach.

2.4 Senior Leaders Now Focus Too Much on the Short Term

- The focus of senior leaders has shifted to become much more about managing short-term performance.
- This has led to more focus on working 'in' the business instead of working 'on' it.

- It also pushes everyone down in the organisation and dilutes accountability.

2.5 Current Business Improvement Models Are Not Working

- Many current improvement initiatives are not delivering sustainable results.
- Efforts are being focussed on quick-win, short-term, discrete solutions, which usually don't last.
- These are not systemic, don't consider the operating system as a whole, and waste a lot of effort.

CHAPTER 3: THE OPERATIONAL BASICS NEED TO BE TAKEN TO A WHOLE NEW LEVEL

3.1 Focus On Optimising the Whole System, Not Just the Parts

- What is needed is an operating model that aligns everyone and everything around delivery of the business strategy.
- This needs to consider not just the value chain but multiple dimensions (geography, hierarchy, time frames, plan/execute/improve, people/process/technology, etc.).
- It should focus not just on the different parts but also on the interfaces and especially the value of the whole system.

3.2 Systemically Improving the Quality of Planning, Execution, and Improvement

- The core of the operating model must be to achieve outstanding planning, execution, and improvement.
- The focus on planning and improvement must be increased, as the main drivers of long-term systemic, sustainable value.
- An enabling physical and organisational working environment needs to be created, which makes it easier for teams to achieve this.

3.3 Fully Integrated Planning

- The planning needs to be well-informed, fully integrated, and optimised for whole-system value.
- The decision-making around the plan assumptions is an important process and requires some independent oversight.
- Tight reconciliation between plan, schedule, execute, and improve is important to ensure the plan remains realistic and on track.

3.4 Disciplined Execution

- A focus on disciplined execution of the plan is essential with a high focus on compliance to plan/schedule.
- The cornerstone tools for this are integrated scheduling and an aligned and cascaded management process.
- A well-resourced process for managing dynamic changes is also key to drive predictability and value optimisation.

3.5 Continuous, Sustainable Improvement

- A change is needed to establish a focus on continuous, systemic, sustainable improvement, not just short-term initiatives.
- This will create a solid platform for a progressive technology strategy.
- Strong governance is required over the new systems and processes to build and underpin a sustainable performance culture.

3.6 Hard-Wiring Safety and Risk into Your Routine Management Processes

- It stands to reason that good planning and disciplined execution are key to a safer operation.
- There are currently many degrees of freedom in frontline operations, and safety is too reliant on culture and standards rather than being rigorously embedded into management processes.

- Similarly, risk management needs to be embedded into the ongoing routine management processes rather than treated as an annual risk-review process.

3.7 Leveraging the Value of Data

- Good data is an important starting point, but information from that data quickly becomes the challenge.
- Once you get good information, you realise that getting people to act on it is easier said than done.
- Creating a systemic and sustainable platform for this rapid decision-making then becomes the ultimate goal.

3.8 Future-Proofing the Changes

- Establishing a 'home' for these changes is important to bring together these different aspects and assist with collaboration and dynamic decision-making.
- The ultimate form for this is an Integrated Operations Centre (IOC), which serves as a nerve centre for the operation.
- IOC design needs to be approached with caution, as the physical facility is less important than the systems, processes, functionality, and mandates behind it (this is often misunderstood by businesses).

CHAPTER 4: INTEGRATION PROVIDES THE MISSING LINK FOR TRANSFORMATIONAL CHANGE

4.1 Integration Is a Systemic Approach, Which Is What Underpins All Great Companies

- Investment in a systemic approach is the key differentiator of truly great companies.
- 'Good' companies are constantly changing, reinventing, and chasing after new initiatives.

- 'Great' companies have a clearly defined operating model and continuously invest in improving the methodology.

4.2 Integration Provides a Structured Framework for a Systemic Operating Model

- A structured but flexible framework built around five key elements is what underpins sustainability.
- It simplifies and aligns all aspects of the operating model through holistic design.
- It hard-wires the operating strategy, creates the right enabling environment, and addresses sustainability and governance.

4.3 Integration Is an Objective, Measurable Approach

- Integration maturity can be objectively measured using an internationally recognised criteria for business excellence.
- Combined with the structured elements in the integration framework, it becomes a very objective and practical methodology to apply.
- This provides a powerful tool for benchmarking against others in the industry and tracking your own status and progress.

4.4 Integration Is an Untapped Area of Huge Latent Value

- Integration maturity-benchmarking data highlights enormous untapped value potential through full integration.
- The gap between mining leaders and laggers is small compared to the value potential through full integration.
- Variability of integration maturity within individual businesses is extremely wide, which is a problem in itself.

4.5 Integration Provides a Solid Platform for a Progressive Technology Strategy

- Any progressive technology or digital strategy needs to be built on a solid operating platform.

- A fully integrated operation provides that platform and positions well for the subsequent 'optimisation' and 'automation' horizons, which will further transform your business.
- This staged approach is essential to transition your business effectively through the transformational change required.

CHAPTER 5: A FIVE-STEP GUIDE FOR INTEGRATING YOUR BUSINESS

5.1 Approach It as a Journey, Not a Big Bang, and Closely Engage Your Leadership

- Systemic, transformational cultural change needs to be managed as a journey, not a big bang.
- Close engagement of the leadership team is essential to bring along the hearts and minds of the whole workforce.
- The leadership team needs to codesign the vision and roadmap to ensure full ownership, understanding, and commitment to the journey.

5.2 Develop a Staged Roadmap, Which Delivers Quick Wins but Also Longer-Term Goals

- A well-structured and carefully designed roadmap significantly accelerates results and reduces risks.
- Staging the journey ensures early gains are built on a solid foundation and will not slip back.
- The integration, optimisation, and automation stages occur in parallel, not in series, but must build progressively.

5.3 Customise This to Align with Your Business Strategy and Existing Initiatives

- If well-designed, an integrated operations (IO) programme should *enhance* any existing initiatives, not conflict with them.
- This whole process will help you to identify and address any

existing gaps and overlaps between activities across the organisation.

- By design, the IO programme will be completely aligned with the overall business strategy.

5.4 Build the Capability of Your Leaders to Lead This and Upskill Your Workforce

- This journey represents a paradigm shift for leaders and the workforce, and they need to be well-prepared.
- Leaders must be trained in the art of leading and managing teams, especially within an integrated mining system, so they are capable of leading the change (practical skills, not theory).
- The workforce needs to be trained to understand the behaviours required to operate within the new operating model.

5.5 Follow a Comprehensive Integrated Operations Framework for Implementation

- A structured, comprehensive approach is essential, using a holistically designed, fully integrated operating model.
- Some full-time internal resourcing will be required, along with a well-managed PMO (programme management office).
- External help with integrated operations expertise is highly recommended, but the transformation process should not be 'farmed out'.

CONCLUSION

- There are enormous benefits to be gained from a fully integrated operating model.
- Any operation that aspires to true operational excellence should follow this approach.

- And any operation that intends to move towards a digital future shouldn't think of doing so without doing this first.
- Time is not on your side; you should move fast.

ACKNOWLEDGMENTS

///

I dedicate this book to my wife, Vera, who sadly passed away halfway through my writing it. Her steadfast support throughout my career, and the encouragement she gave me to write this book, are a shining demonstration of the value and importance of family in all that we do in life. If this book brings a little more clarity and purpose to your working life, and removes some of the frustration, clutter, and confusion in your head, it will have been worthwhile.

To all of my colleagues at NextGenOpX, I am inspired by the quality and dedication of our team, and the passion you all share with me for fixing some deeply rooted paradigms that exist in the mining industry. It has been a great journey to date and we are on a good path to changing the industry for the better. You have all contributed to the content on these pages. Thank you to Shane Johnson, Lionel Louw, Graeme Stanway, Gideon Malherbe, Chris Taylor, Werner Grundling, Andrew Wildy, Dean Hoare,

Ernst Griebel, Stephen Sherring, and Geoff Day for your support and guidance on this journey and your review of my early drafts.

And finally, thank you to the Scribe Media team for helping me polish the quality of this final product. Special thanks to Erin Mellor for steering me through the whole process, Nicole Jobe for your excellent help with the editing, Skyler Gray on the title selection, and Joel Gendron for the cover design.

I have learnt that writing a book is not for the faint-hearted. However, it is a deeply satisfying thing, and a privilege, to have the opportunity to articulate one's thoughts in such an organised way. I hope those who read it will find it brings value and inspiration.